REAPING THE BENEFITS OF Genomic and Proteomic Research

Intellectual Property Rights, Innovation, and Public Health

Committee on Intellectual Property Rights in Genomic and Protein Research and Innovation

Board on Science, Technology, and Economic Policy

Committee on Science, Technology, and Law

Policy and Global Affairs

Stephen A. Merrill and Anne-Marie Mazza, Editors

NATIONAL RESEARCH COUNCIL
OF THE NATIONAL ACADEMIES

THE NATIONAL ACADEMIES PRESS
Washington, D.C.
www.nap.edu

THE NATIONAL ACADEMIES PRESS 500 Fifth Street, N.W. Washington, DC 20001

NOTICE: The project that is the subject of this report was approved by the Governing Board of the National Research Council, whose members are drawn from the councils of the National Academy of Sciences, the National Academy of Engineering, and the Institute of Medicine. The members of the committee responsible for the report were chosen for their special competences and with regard for appropriate balance.

This study was supported by Contract No. N01-OD-4-2139, between the National Academies and the National Institutes of Health. In addition, the study was supported by Affymetrix Inc., Agilent Technologies, Amgen, Biotechnology Industry Organization, Chiron Foundation, Pfizer Inc., and the Bellagio Center of the Rockefeller Foundation. The views presented in this report are those of the National Research Council Committee on the Intellectual Property Rights in Genomic and Protein Research and Innovation and are not necessarily those of the funding agencies.

International Standard Book Number 0-309-10067-4 (Book)

International Standard Book Number 0-309-65523-4 (PDF)

Library of Congress Control Number 2006921870

Additional copies of this report are available from the National Academies Press, 500 Fifth Street, N.W., Lockbox 285, Washington, DC 20055; (800) 624-6242 or (202) 334-3313 (in the Washington metropolitan area); Internet, http://www.nap.edu.

Printed in the United States of America.

THE NATIONAL ACADEMIES

Advisers to the Nation on Science, Engineering, and Medicine

The **National Academy of Sciences** is a private, nonprofit, self-perpetuating society of distinguished scholars engaged in scientific and engineering research, dedicated to the furtherance of science and technology and to their use for the general welfare. Upon the authority of the charter granted to it by the Congress in 1863, the Academy has a mandate that requires it to advise the federal government on scientific and technical matters. Dr. Ralph J. Cicerone is president of the National Academy of Sciences.

The **National Academy of Engineering** was established in 1964, under the charter of the National Academy of Sciences, as a parallel organization of outstanding engineers. It is autonomous in its administration and in the selection of its members, sharing with the National Academy of Sciences the responsibility for advising the federal government. The National Academy of Engineering also sponsors engineering programs aimed at meeting national needs, encourages education and research, and recognizes the superior achievements of engineers. Dr. Wm. A. Wulf is president of the National Academy of Engineering.

The **Institute of Medicine** was established in 1970 by the National Academy of Sciences to secure the services of eminent members of appropriate professions in the examination of policy matters pertaining to the health of the public. The Institute acts under the responsibility given to the National Academy of Sciences by its congressional charter to be an adviser to the federal government and, upon its own initiative, to identify issues of medical care, research, and education. Dr. Harvey V. Fineberg is president of the Institute of Medicine.

The **National Research Council** was organized by the National Academy of Sciences in 1916 to associate the broad community of science and technology with the Academy's purposes of furthering knowledge and advising the federal government. Functioning in accordance with general policies determined by the Academy, the Council has become the principal operating agency of both the National Academy of Sciences and the National Academy of Engineering in providing services to the government, the public, and the scientific and engineering communities. The Council is administered jointly by both Academies and the Institute of Medicine. Dr. Ralph J. Cicerone and Dr. Wm. A. Wulf are chair and vice chair, respectively, of the National Research Council.

www.national-academies.org

COMMITTEE ON INTELLECTUAL PROPERTY RIGHTS IN GENOMIC AND PROTEIN RESEARCH AND INNOVATION

SHIRLEY TILGHMAN (*Co-Chair*), President, Princeton University, Princeton, NJ
RODERICK MCKELVIE (*Co-Chair*), Partner, Covington & Burling, Washington, DC
ASHISH ARORA, Professor, H. John Heinz III School of Public Management, Carnegie Mellon University, Pittsburgh, PA
HELEN M. BERMAN, Board of Governors Professor of Chemistry and Chemical Biology, Rutgers University, Piscataway, NJ
JOYCE BRINTON, former Director, Office for Technology and Trademark Licensing, Harvard University, Cambridge, MA
STEPHEN BURLEY, Chief Scientific Officer, SGX Pharmaceuticals, Inc., San Diego, CA
Q. TODD DICKINSON, Vice President and Chief Intellectual Property Counsel, General Electric Co., Fairfield, CT
ROCHELLE DREYFUSS, Pauline Newman Professor of Law, New York University School of Law, New York, NY
REBECCA S. EISENBERG, Robert and Barbara Luciano Professor of Law, University of Michigan Law School, Ann Arbor, MI
CHARLES HARTMAN,[1] General Partner, CW Ventures, New York, NY
DANIEL KEVLES, Stanley Woodward Professor of History, Yale University, New Haven, CT
DAVID KORN, Senior Vice President, Biomedical and Health Sciences Research, Association of American Medical Colleges, Washington, DC
GEORGE MILNE, JR., Partner, Radius Ventures, New York, NY
RICHARD SCHELLER, Executive Vice President, Research, Genentech, Inc., San Francisco, CA
ROCHELLE SEIDE, Partner, Arent Fox PLLC, New York, NY
ROBERT WATERSTON, Professor and Gates Chair, Genome Sciences, University of Washington, Seattle, WA
NANCY WEXLER, Higgins Professor Neuropsychology, College of Physicians and Surgeons of Columbia University and President, Hereditary Disease Foundation, New York, NY
BRIAN WRIGHT, Professor of Agricultural and Resource Economics, University of California, Berkeley, Berkeley, CA

[1]Deceased 2/22/05

Staff

STEPHEN A. MERRILL, Study Director
ANNE-MARIE MAZZA, Study Director
KATHI E. HANNA, Consultant
SARA DAVIDSON MADDOX, Editor
LAURA SHEAHAN, Program Officer
CRAIG SCHULTZ, Senior Research Associate (through August 2005)
PATRICIA E. SANTOS, Program Associate (through August 2005)
STACEY E. SPEER, Project Associate (through March 2005)
ROBERT BERLIN, Christine Mirzayan Science and Technology Fellow (Summer 2005)
KAREN BOYD, Christine Mirzayan Science and Technology Fellow (Winter 2005)
SARAH ELSON, Christine Mirzayan Science and Technology Fellow (Summer 2005)
MARA JEFFRESS, Christine Mirzayan Science and Technology Fellow (Winter 2005)
PETER KOZEL, Christine Mirzayan Science and Technology Fellow (Winter 2004)
SARA LIGHTBODY, Christine Mirzayan Science and Technology Fellow (Summer 2004)
ANNA STAVLA, Christine Mirzayan Science and Technology Fellow (Fall 2004)
ELENI ZIKA, Christine Mirzayan Science and Technology Fellow (Summer 2004)

COMMITTEE ON SCIENCE, TECHNOLOGY, AND LAW

DONALD KENNEDY (NAS/IOM) (Co-Chair), Editor in Chief, Science, President Emeritus and Bing Professor of Environmental Science Emeritus, Stanford University
RICHARD A. MERRILL (IOM), (Co-Chair), Daniel Caplin Professor of Law, University of Virginia Law School
SHIRLEY S. ABRAHAMSON, Chief Justice, Wisconsin Supreme Court
FREDERICK R. ANDERSON, JR., Partner, McKenna, Long & Aldridge LLP
MARGARET A. BERGER, Suzanne J. and Norman Miles Professor of Law, Brooklyn Law School
ARTHUR I. BIENENSTOCK, Vice Provost and Dean of Research and Graduate Policy, Stanford University
PAUL D. CARRINGTON, Professor of Law, Duke University Law School
JOE S. CECIL, Project Director, Program on Scientific and Technical Evidence, Division of Research, Federal Judicial Center
JOEL E. COHEN (NAS), Abby Rockefeller Mauze Professor and Head, Laboratory of Populations, The Rockefeller University and Columbia University
KENNETH W. DAM, Max Pam Professor Emeritus of American and Foreign Law and Senior Lecturer, University of Chicago Law School
REBECCA S. EISENBERG, Robert and Barbara Luciano Professor of Law, University of Michigan Law School
DAVID J. GALAS, Chancellor and Chief Scientific Officer, Norris Professor of Applied Life Sciences, Keck Graduate Institute of Applied Life Sciences
DAVID L. GOODSTEIN, Vice Provost; Professor of Physics and Applied Physics; Frank J. Gilloon Distinguished Teaching and Service Professor, California Institute of Technology
SHEILA S. JASANOFF, Pforzheimer Professor of Science and Public Policy Studies, John F. Kennedy School of Government, Harvard University
DANIEL J. KEVLES, Stanley Woodward Professor of History, Yale University
DAVID KORN (IOM), Senior Vice President for Biomedical and Health Sciences Research, Association of American Medical Colleges
ROBERT A. LONERGAN, Vice President and General Counsel, Rohm and Haas
PATRICK A. MALONE, Partner, Stein, Mitchell & Mezines
RICHARD A. MESERVE (NAE), President, Carnegie Institution of Washington
ALAN B. MORRISON, Senior Lecturer, Stanford Law School

THOMAS D. POLLARD (NAS/IOM), Eugene Higgins Professor, Department of Molecular, Cellular, and Developmental Biology, Yale University
CHANNING R. ROBERTSON, Ruth G. and William K. Bowes Professor, Dean of Faculty and Academic Affairs, School of Engineering, Stanford University
JONATHAN M. SAMET (IOM), Professor and Chairman, Department of Epidemiology, Johns Hopkins Bloomberg School of Public Health
FERN M. SMITH, U.S. District Judge (retired), U.S. District Court for the Northern District of California
JAMES GUSTAVE SPETH, Dean and Professor in the Practice of Environmental Policy and Sustainable Development, Yale School of Forestry and Environmental Studies
SHEILA E. WIDNALL (NAE), Institute Professor, Massachusetts Institute of Technology

Staff

ANNE-MARIE MAZZA, Director
ELIZABETH BRIGGS-HUTHNANCE, Senior Program Associate
PATRICIA E. SANTOS, Program Associate (through August 2005)
STACY SPEER, Program Associate (through March 2005)

Preface and Acknowledgments

The free exchange and open dissemination of scientific information and the pursuit of basic scientific research has led to remarkable advances in our understanding of biology. This scientific progress, coupled with the protections and information dissemination possibilities offered by a vigorous patent system, has resulted in the development of numerous products—with many more on the horizon—that can be used to diagnose, treat, and cure a variety of diseases.

Avoiding a conflict between open dissemination and access to scientific discoveries and the protection of inventors' rights is critical to furthering scientific progress and enhancing human health. It also is critical that as science evolves, we stop to assess whether the appropriate mechanisms to prevent such a conflict remain in place. This report is just such an assessment—a marker in time that looks at the state of genomic and proteomic research and the current policies and practices promoting or restricting the dissemination of scientific information, tools, and products, and asks, "are there any storms over the horizon?"

The original survey data collected for this report, although arguably the best data currently available to address some of the committee's questions, necessarily reflect a limited snapshot of the current situation. This survey produced some important findings, which the committee took into account in deliberating on its recommendations. Yet in light of a modest response rate and other limitations inherent in survey research, the committee also drew from many other sources of information in addition to its own research, including the presentations of many speakers at its public meetings and workshops, interviews conducted with academic and industry representatives, and informal discussions with colleagues at its members' own institutions. The report focuses more deliberately on genomics because it has a longer scientific and intellectual property history, while proteomics is still in a nascent stage, with patenting activities just gearing up.

Based on these many sources of information, the committee offers recommendations intended to foster scientific advances, maintain a strong patent system, and create an environment in which significant contributions can be made to scientific progress and human health. The committee calls on scientists in all sectors to incorporate the norms espoused in this report and urges those who interpret and enforce the patent system to be mindful of the evolving nature of science and the ways in which it continues to challenge our assumptions.

The committee gratefully acknowledges the contributions of the following individuals who provided invaluable information and other assistance. Affiliations are at the time of participation in the study: **Elena Armandola**, European Patent Office; **Robert Armitage**, Eli Lilly and Company; **Alan Bennett**, University of California, Davis; **Jeremy Berg**, National Institutes of Health; **David Beyer**, Amgen; **Sara Boettiger**, University of California, Berkley; **Charles Caruso**, Merck & Co., Inc; **Scott Chambers**, Patton Boggs LLP; **Charlene Cho**, University of Chicago; **Iain Cockburn**, Boston University; **Wesley Cohen**, Duke University; **Francis Collins**, National Human Genome Research Institute; **Robert Cook-Deegan**, Duke University; **Kenneth Dam**, University of Chicago Law School; **Dennis Drayna**, National Institutes of Health; **Anthony Delcampo**, Dana-Farber Cancer Institute; **Susan Ehringhaus**, Association for American Medical Colleges; **George Elliott**, U.S. Patent and Trademark Office; **Lila Feisse**, BIO; **Joanna Groden**, University of Cincinnati; **Corey Goodman**, Renovis; **Stephen Hansen**, American Association for the Advancement of Science; **Udo Heinemann**, Max Delbrueck Centrum fur Molekulare Medizin; **Brian Hicks**, Brigham and Women's Hospital; **Stephen Hilgartner**, Cornell University; **Hans-Rainer Jaenichen**, Vossius & Partner; **Ester Kepplinger**, U.S. Patent and Trademark Office; **Hiroki Kitamura**, Japanese Patent Office; **Robert Kneller**, University of Tokyo; **Katharine Ku**, Stanford University; **Stephen Kunin**, Oblon & Spivak; **Jeffrey Kushan**, Sidly, Austin, Brown & Wood LLP; **Waldemar Kutt**, DG Research, European Commission; **Debra Leonard**, Weill Medical College of Cornell University; **Tim Leshan**, National Human Genome Research Institute; **Gert Matthijs**, Center for Human Genetics; **Jon Mertz**, University of Pennsylvania; **Arie Michelsohn**, Finnegan, Henderson, Farabow, Garrett & Dunner; **Sadao Nagaoka**, Hitotsubashi University; **Pauline Newman**, U.S. Court of Appeals for the Federal Circuit; **John Norvell**, National Institute of General Medical Sciences; **Frank Orlaudella**, Agilent; **Marvin Parnes**, University of Michigan; **Lori Pressman**, Consultant; **Richard Rodriquez**, National Institutes of Health; **Mark Rohrbaugh**, National Institutes of Health; **Tony Rollins**, Merck & Co., Inc; **Bill Rusconi**, Myriad Genetic Laboratories; **Christina Sampogna**, Organization for Economic Cooperation and Development; **Mark Sobel**, Association for Molecular Pathology; **Jon Soderstrom**, Yale University; **Melissa Soucy**, Georgetown University; **Avron Spier**, Allon Therapeutics, Inc.; **Ashley**

Stevens, Boston University; **Edwin Stone**, University of Iowa; **John Sulston**, Wellcome Trust Sanger Institute; **Koichi Sumikura**, National Graduate Institute for Policy Studies (Japan); **Lawrence Sung**, University of Maryland; **Akiteru Tamura**, Japanese Patent Office; **Fred Telling**, Pfizer; **Sandy Thomas**, Nuffield Bioethics Council; **Sara Vinarov**, Quarles & Brady LLP; **Shawna Vogel**, Massachusetts Institute of Technology; **John Walsh**, University of Illinois, Chicago; **LeRoy Walters**, Georgetown University; and **Robert Wells**, Affymetrix.

This report has been reviewed in draft form by individuals chosen for their diverse perspectives and technical expertise, in accordance with procedures approved by the National Academies' Report Review Committee. The purpose of this independent review is to provide candid and critical comments that will assist the institution in making its published report as sound as possible and to ensure that the report meets institutional standards for objectivity, evidence, and responsiveness to the study charge. The review comments and draft manuscript remain confidential to protect the integrity of the process.

We wish to thank the following individuals for their review of this report: **Robert Armitage**, Eli Lilly and Company; **Wendy Baldwin**, University of Kentucky; **Fred Cohen**, University of California, San Francisco; **Robert Cook-Deegan**, Duke University; **Deborah Delmer**, Rockefeller Foundation; **Anthony Ford-Hutchinson**, Merck Research Laboratories; **Steven Holtzman**, Infinity Pharmaceuticals, Inc.; **Stephen Kunin**, Oblon & Spivak; **Lonnie Ingram**, University of Florida; **Joshua LaBaer**, Harvard University; **Gilbert Omenn**, University of Michigan; **Arti Rai**, Duke University; **James Severson**, University of Washington; **Sandy Thomas**, Nuffield Council on Bioethics; **Marie Thursby**, Georgia Tech; **Phillip Sharp**, Massachusetts Institute of Technology; **Leroy Walters**, Georgetown University; **Charles Wilson**, Consultant; and **Barbara Zehnbauer**, Washington University in St. Louis.

Although the reviewers listed above have provided many constructive comments and suggestions, they were not asked to endorse the conclusions or recommendations, nor did they see the final draft of the report before its release. The review of this report was overseen by **John Bailar**, University of Chicago, and **Kenneth Dam**, University of Chicago. Appointed by the National Academies, they were responsible for making certain that an independent examination of this report was carried out in accordance with institutional procedures and that all review comments were carefully considered. Responsibility for the final content of this report rests entirely with the authoring committee and the institution.

We are indebted to the input of the entire committee and thank each member for his or her willingness to wrestle with these complex issues and to put forth what the committee agrees is the best public policy for genomic and proteomic research. We also wish to thank staff and consultants—Stephen Merrill, Anne-Marie Mazza, Kathi Hanna, Craig Schultz, Sara Davidson Maddox, Stacey Speer, Patricia Santos, and a succession of participants in the National Research Council's Christine Mirzayan Science and Technology Policy Fellowship Program—for their assistance and dedication. We are also indebted to Tim Leshan, Senior Policy Analayst, National Human Genome Research Institute, for his extremely effective liaison between the committee and all of the interested parties at NIH, and to the Rockefeller Foundation's Bellagio Center for providing an ideal venue in which to explore similarities and differences in approaches to these issues with European and Japanese colleagues. We thank John Walsh and Charlene Cho, University of Illinois, Chicago, and Wesley Cohen, Duke University, for developing and conducting the survey of research scientists that added much to our understanding of intellectual property from the perspective of the biomedical research bench.

Finally, we wish to acknowledge one committee member in particular, Charles (Chuck) Hartman. Chuck died suddenly on February 22, 2005, and did not see the final report, although his contributions throughout the committee deliberations helped shape its outcome in innumerable ways. We hope in some way that the report reflects his graciousness, thoughtfulness, and commitment to sound public policy.

Shirley Tilghman and Roderick McKelvie

Contents

List of Boxes, Figures, and Tables

BOXES

FIGURES

TABLES

Summary

The nature of biological inquiry and the norms of behavior in the scientific community have changed in the wake of the Human Genome Project (HGP) and the birth of proteomics. Complementing the traditional hypothesis-driven study of single genes or proteins is the option of studying many genes or proteins simultaneously. This sea change has occurred while both universities and industry have been aggressively seeking and defending intellectual property protection for discoveries, many of them well upstream of commercial application. Thus the potential for a "perfect storm" exists, in which future discoveries in genomics and proteomics that would benefit the public health and well-being could be thwarted by an increasingly complex intellectual property regime.

In light of this changing environment, the National Institutes of Health (NIH) asked the National Academy of Sciences to study the granting and licensing of intellectual property rights on discoveries relating to genetics and proteomics and the effects of these practices on research and innovation. Specifically, NIH asked the National Academy to study and report on:

1. trends in the number and nature of U.S.-issued patents granted for technologies related to genomics and proteomics;
2. standards that the U.S. Patent and Trademark Office (USPTO) and other patent offices (specifically in Europe and Japan) are applying in acting on these applications;
3. the effects of patenting genomic and proteomic inventions and/or licensing practices for inventions on research and innovation; and

4. steps that NIH and others might take to ensure the productivity of research and innovation involving genes and proteins.

Under the auspices of NAS's Science, Technology, and Economic Policy Board and Committee on Science, Technology, and Law, a study committee was formed in response to this charge. The committee reviewed the literature in the field, held several public sessions that included presentations by experts and stakeholders, and conducted a survey of how biomedical research scientists acquire, use, and experience intellectual property practices. Based on these sources of information, the committee drew conclusions and made recommendations in three broad areas that aim to ensure that the public investment in genomics and proteomics results in optimal public benefit:

1. improving and facilitating best practices and norms in the conduct of genomics and proteomic research:
2. adapting the patent system to the rapidly changing fields of genomics and proteomics; and
3. facilitating research access to patented inventions through licensing and shielding from liability for infringement.

The committee was not asked to and did not directly address issues in the acquisition and management of and access to intellectual property involving plants and animals. Clearly, differences in the scale and diversity of research and in the distribution of patent ownership between agricultural and human biotechnology exist, and these variations may merit separate study. Nevertheless, there has been comparable progress in plant molecular biology. Gene-based diagnostics also are important for monitoring crop and farm animal diseases. All of these have been subjects of patenting in the same manner as human DNA sequences, genes, and proteins. The committee suggests that, when these similarities warrant, it may be appropriate to apply to animal and plant biotechnology comparable policies and principles to those that it recommends for biomedical research and development.

OVERALL CONCLUSIONS

The committee found that the number of projects abandoned or delayed as a result of difficulties in technology access is reported to be small, as is the number of occasions in which investigators revise their protocols to avoid intellectual property issues or in which they pay high costs to obtain intellectual property. Thus, for the time being, it appears that access to patented inventions or information inputs into biomedical research rarely imposes a significant burden for biomedical researchers. For a number of reasons, however, the committee concluded that the patent landscape, which already is becoming complicated in areas such as gene expression and protein-protein interactions, could become considerably more complex and burdensome over time.

There are several reasons to be cautious about the future. The lack of substantial evidence for a patent thicket or a patent-blocking problem is associated with a general lack of awareness or concern among academic investigators about existing intellectual property. That could change dramatically and possibly even abruptly under two circumstances. First, institutions, as they become aware that they may enjoy no protection from legal liability,[1] may become more concerned about their potential patent infringement liability and take more active steps to raise researchers' awareness or even to try to regulate their behavior. The latter could be both burdensome on research *and* largely ineffective because of the autonomy of academic researchers and their ignorance—or at best uncertainty—about what intellectual property laws apply in what circumstances. Alternatively, patent holders, aware that universities are not especially shielded by law from patent infringement liability, could take more active steps to assert their competing patents. This may not lead to more patent suits against universities or between companies—indeed, established companies usually are reluctant to pursue litigation against research universities—but it could involve more demands for licensing fees, grant-back rights, and other terms that are burdensome to research. Certainly, some holders of gene-based diagnostic patents currently are active in asserting their intellectual property rights.

Finally, as scientists increasingly use the high-throughput tools of genomics and proteomics to study the properties of many genes or proteins simultaneously, the burden on the investigator to obtain rights to the intellectual property could become insupportable.

Perhaps most importantly, the results of the survey conducted with the support of the committee revealed substantial evidence of a more immediate and potentially remediable burden on research—private as well as public—stemming from difficulties in accessing proprietary research materials, whether patented or unpatented. The committee found that impediments to the exchange of biomedical research materials remain prevalent and may be increasing.

Several steps may be taken to anticipate and prevent the emergence of an

[1]The situation for state institutions is more complex and may provide state institutions with greater protection. Under the 11th Amendment states enjoy immunity from suits in federal courts for monetary damages absent either their express consent or a legitimate congressional grant of the power to sue in federal court. Although a state can be enjoined from continuing to infringe, Congress only can abrogate 11th Amendment immunity if it is remedying a failure by the states to provide adequate compensation for unauthorized past usage, in this case, the failure to provide legitimate protection of rights through the state's courts. To be deemed adequate the state courts must, however, provide both due process for petitioners and the possibility of redress.

increasingly problematic environment for research in genomics in the near future and for proteomics further out, as more knowledge is created, more patent applications are filed, and more restrictions might be placed on the availability of and access to information and resources.

BEST PRACTICES AND NORMS FOR THE SCIENTIFIC COMMUNITY AND FEDERAL RESEARCH SPONSORS

Many of the potential problems looming in the realm of genomics, proteomics, and intellectual property can be avoided if scientists and their institutions, whether public or private, follow the best practices already articulated by the National Institutes of Health (NIH), the National Research Council (NRC), and others. U.S. science has flourished because of its general openness and the sharing of data and research resources. This is not to suggest that legitimate proprietary interests in science do not exist, but rather is intended to highlight the argument that whenever possible, sharing is in the best interest of all science, both basic and applied. Several measures can be taken to facilitate the free exchange of materials and data.

Foster Free Exchange of Data, Information, and Materials

From the inception of the HGP, public and commercial funders of these activities have emphasized that, in order to reap the maximum benefit to the public health, data should be freely available in the public domain. In addition, the NRC has repeatedly emphasized the need for sharing data. The council's 2003 report *Sharing Publication-Related Data and Materials* endorsed the uniform principle for sharing integral data and materials expeditiously:

> Community standards for sharing publication-related data and materials should flow from the general principle that the publication of scientific information is intended to move science forward. More specifically, the act of publishing is a *quid pro quo* in which authors receive credit and acknowledgement in exchange for disclosure of their scientific findings. An author's obligation is not only to release data and materials to enable others to verify or replicate published findings but also to provide them in a form on which other scientists can build with further research. All members of the scientific community—whether working in academia, government, or a commercial enterprise—have equal responsibility for upholding community standards as participants in the publication system, and all should be equally able to derive benefits from it (NRC, 2003, p. 4).

Nucleic acid sequences provide the fundamental starting point for describing and understanding the structure, function, and development of genetically diverse organisms. For almost 20 years, GenBank, the European Molecular Biology Laboratory, and the DNA Data Bank of Japan have collaborated to create nucleic acid sequence data banks. These data banks are invaluable to researchers but they face insufficiencies and gaps as fewer data deposits are made because of proprietary interests.

The genomics and proteomics communities, in general, have honored these calls for data sharing, especially in the large-scale projects such as the HGP itself, the Expressed Sequence Tag (EST) project, and the SNP Consortium. Some practices, however, do not conform to these norms. Private industry consistently retains some portion of its protein structure information in proprietary databases, effectively withholding from the scientific community a large and important dataset that could facilitate basic and applied research in structural biology. However, once structures are no longer commercially important, their availability in the public domain would be very useful for academic research. In addition, defensive patenting of three-dimensional structures of drug targets has the potential to interfere with drug discovery. Structural biology data in published patent applications and issued patents are presented in such a form that they are not readily incorporated into the Protein Data Bank (PDB) for the benefit of the larger scientific community. Furthermore, academic scientists are sometimes driven by competitive pressures to withhold both information and materials.

Eventually, large-scale structural genomics efforts will dominate the production of new structures. Full disclosure of structures without patenting could serve to preempt much of the defensive patenting currently sought by industry and substantially improve the environment for all of science. The committee commends NIH for its effective use of provisions in Requests for Proposals for projects involving the development of resources for the public domain that require that grant applicants include in their proposals an explanation of their plans for the sharing and dissemination of research results. Although NIH does not currently collect and analyze data on grantee behavior, it has the ability and the authority to elicit good behavior among grantees and contractors and should exercise that authority wherever possible.

Recommendation 1:
NIH should continue to encourage the free exchange of materials and data. NIH should monitor the actions of grantees and contractors with regard to data and material sharing and, if necessary, require grantees and contractors to comply with their approved intellectual property and data sharing plans.

However, it should be noted that investigators have the right and even the obligation to retain materials and data until they are confident of their validity and have reported their results in publication. The quality of science and the value of the public data must be upheld even while meeting the goal of sharing materials and data.

Recommendation 2:
The committee supports NIH in its efforts to adapt and extend the "Bermuda Rules" to structural biology data generated by NIH-funded centers for large-scale structural genomics efforts and thereby make data promptly and freely available in a database via the PDB.

Although in principle the coordinate data that are in patent applications could be put into the PDB, both the content and format of these patent applications are not suitable for incorporation into the repository. The PDB has established standard formats for electronically archiving the coordinate, experimental, and meta data. Recently USPTO proposed that these data be sent in electronic form as part of relevant patent applications. The Worldwide PDB, an international organization responsible for all PDB data, endorsed this proposal and further stipulated that the standard formats be required. This would ensure that the data would be efficiently and properly archived and be made freely available.

Recommendation 3:
The PDB should work with USPTO, the European Patent Office (EPO), and the Japanese Patent Office (JPO) to establish mechanisms for the efficient transfer of structural biology data in published patent applications and issued patents to the PDB for the benefit of the larger scientific community. To the extent feasible within commercial constraints, all researchers, including those in the private sector, should be encouraged to submit their sequence data to GenBank, the DNA Databank of Japan, or the European Molecular Biology Laboratory and to submit their protein structure data to the PDB.

Foster Responsible Patenting and Licensing Strategies

In 1999, NIH issued *Principles and Guidelines for Recipients of NIH Research Grants and Contracts on Obtaining and Disseminating Biomedical Research Resources* (64 FR 72090).[2] These aspirational principles were issued by NIH to provide guidance and direction to NIH-funded institutions in order to balance the need to protect intellectual property rights with the need to broadly disseminate new discoveries. They recognize that licensing policies and practices are extremely important determinants of the effects of patents on upstream technologies on the conduct of follow-on research. The principles apply to all NIH-funded entities and address biomedical materials, which are broadly defined to include cell lines, monoclonal antibodies, reagents, animal models, combinatorial

[2]A copy of the complete principles can be obtained at the NIH Web site at *http://www.nih.gov/od/ott/RTguide_final.htm*.

chemistry libraries, clones and cloning tools, databases, and software (under some circumstances).[3]

The principles were developed in response to complaints from researchers that restrictive terms in material transfer agreements (MTAs) were impeding the sharing of research resources. These restrictions came both from industry sponsors and from research institutions. In the *Principles and Guidelines*, NIH urges recipient institutions to adopt policies and procedures to encourage the exchange of research tools by minimizing administrative impediments, ensuring timely disclosure of research findings, ensuring appropriate implementation of the Bayh-Dole Act, and ensuring the dissemination of research resources developed with NIH funds.

Consistent with its ongoing interest in facilitating broad access to government-sponsored research results, NIH in 2004 issued *Best Practices for the Licensing of Genomic Inventions*. This document aims to maximize the public benefit whenever technologies owned or funded by the Public Heath Service are transferred to the commercial sector. In this document, NIH recommends that "whenever possible, non-exclusive licensing should be pursued as a best practice. A non-exclusive licensing approach favors and facilitates making broad enabling technologies and research uses of inventions widely available and accessible to the scientific community." The document goes on to say that "exclusive licenses should be appropriately tailored to ensure expeditious development of as many aspects of the technology as possible." The policy distinguishes between diagnostic and therapeutic applications and cautions against exclusive licensing practices in some areas. For example, the document states that "patent claims to gene sequences could be licensed exclusively in a limited field of use drawn to development of antisense molecules in therapeutic protocols. Independent of such exclusive consideration, the same intellectual property rights could be licensed non-exclusively for diagnostic testing or as a research probe to study gene expression under varying physiological conditions."[4]

The committee endorses these NIH policies, in particular the principles that patent recipients should analyze whether further research, development, and private investment are needed to realize the usefulness of their research results and that proprietary or exclusive means of dissemination should be pursued only when there is a compelling need. Also, whenever possible, licenses should be limited to relatively narrow and specific commercial application rather than as blanket exclusive licenses for uses that cannot be anticipated at the moment.

---

[3]The guidelines were issued following recommendations made to the NIH Advisory Committee to the Director by a special subcommittee chaired by Rebecca Eisenberg.

[4]On April 11, 2005, NIH published the final notice, after receipt of public comments, at *http://ott.od.nih.gov/lic_gen_inv_FR.html.*

Recommendation 4:
The committee endorses NIH's *Principles and Guidelines for Recipients of NIH Research Grants and Contracts on Obtaining and Disseminating Biomedical Research Resources* and *Best Practices for the Licensing of Genomic Inventions*. Through its *Guide for Grants and Contracts*, NIH should require that recipients of all research grant and career development award mechanisms, cooperative agreements, contracts, institutional and Individual National Research Service Awards, as well as NIH intramural research studies, adhere to and comply with these guidance documents. Other funding organizations (such as other federal agencies, nonprofit and for-profit sponsors) should adopt similar guidelines.

These principles can and should be followed by other funding agencies. In addition, they should be followed consistently for gene patents and licenses, and they should be applied to proteomics research to discourage inappropriate patenting and licensing practices. For example, the committee believes that it would be consistent with the NIH guidelines to discourage grantees and contractors from patenting three-dimensional macromolecular structures. For the sake of clarity, the committee does not believe that NIH grantees and contractors should be discouraged from patenting biological macromolecules that have been shown to have clear therapeutic value in their own right. The committee recognizes the value of patents when follow-on private investment adds social value by bringing products and services to market, and while this is to be commended, licensing should be done in ways that permit continued research and avoid logjams, undue royalty stacking, and anti-commons problems.

Because NIH issued these policies as guidance documents, grantees and contractors are not required to comply with them. Nor are researchers and research institutions not funded by NIH under any obligation to comply. The committee believes that NIH should continue to encourage adherence to these guidelines and best practices by the extramural community. However, in circumstances in which grantees are found to be ignoring the guidelines and thereby inhibiting innovation, the committee believes that NIH should use its authority to make adherence to the guidelines a condition of a future grant or contract award. By placing the responsibility with the applicant, NIH can state a position relative to its overall goal, but not generate endless pages of detailed policies and procedures. This is an evolving area where flexibility is important. If the goal is normative behavior, some process must be in place to make institutions and investigators examine their own behavior and articulate how they will behave in the broad context of agreed-upon goals. If those positions are widely shared, as in the grant application process, they will help to develop consensus about acceptable or desirable behavior. If there is flexibility in how institutions can approach these issues, then the entire field will reap the benefit of creative approaches.

In addition, NIH and the broader research community should encourage, wherever possible, voluntary compliance with the intent of these policy documents. There are many precedents for voluntary compliance with such standards by industry, dating back to the voluntary submission of research protocols involving recombinant DNA, and more recently, gene transfer studies, to NIH's Recombinant DNA Advisory Committee for review.

Furthermore, the committee's research found that most institutions report that they reserve rights for their own investigators to use a patented technology even though it is licensed exclusively to a commercial entity. An increasingly common university practice in recent years is to reserve such rights for investigators at other nonprofit institutions, but this often is subject to the patent holder's case-by-case approval. The committee commends and endorses this practice, which could be applied to other organizations, as appropriate.

Recommendation 5:
Universities should adopt the emerging practice of retaining in their license agreements the authority to disseminate their research materials to other research institutions and to permit those institutions to use patented technology in their nonprofit activities.

In addition, to support the dissemination of biological research materials to the scientific research community, institutions use Material Transfer Agreements (MTAs) in handling the exchange of research materials with the research community. MTAs are intended to protect the institution's ownership interest in the research material and contain provisions regarding the distribution and use of the research material. However, in the committee's opinion, the use and complexity of these agreements have become burdensome and overly restrictive. Institutions should promote the exchange of material and data while protecting legitimate intellectual property interests.

Recommendation 6:
In cases in which agreements are needed for the exchange of research materials and/or data among nonprofit institutions, researchers and their institutions should recognize restrictions and aim to simplify and standardize the exchange process. Agreements such as the Simple Letter Agreement for the Transfer of Materials or the Uniform Biological Material Transfer Agreement (UBMTA) can facilitate streamlined exchanges. In addition, NIH should adapt the UBMTA to create a similar standardized agreement for the exchange of data. Industry is encouraged to adopt similar exchange practices.

ADAPTING THE PATENT SYSTEM TO THE DEVELOPING FIELDS OF GENOMICS AND PROTEOMICS

To obtain a patent an applicant must claim an invention that falls within patent-eligible subject matter. The invention must be new, useful, and nonobvious in light of the prior art. The patent application must satisfy certain disclosure requirements, including a written description of the invention, an enabling disclosure that allows a person of ordinary skill in the field to make and use the invention without undue experimentation, and disclosure of the best mode contemplated by the inventor of carrying out the invention. The exclusion of abstract ideas from patent protection traditionally has been more important for information technology than for biotechnology, but some genomics and proteomics research has the potential to confuse or even to blur the boundaries between abstract ideas and applications.

The fields of genomics and proteomics are dependent on rapidly changing technology and complex theory. Understanding biological processes through the association of genetic variation with individual phenotypic differences and through structural analyses will involve a variety of methods (global medical sequencing and population genetics in the first and X-ray crystallography and nuclear magnetic resonance [NMR] spectroscopy in the second). These methods will raise many new challenges for USPTO and the courts.

The challenge of these types of innovations clearly was illustrated in the 1990s when the scientific community was in intense discussions with USPTO about the value of ESTs. It will be increasingly important for patent examiners to be current with scientific and clinical developments in the field.

> ***Recommendation 7:***
> **USPTO should create a regular, formal mechanism, such as a chartered advisory committee or a regularly scheduled forum, comprising leading scientists in relevant emerging fields, to inform examiners about new developments and research directions in their field. NIH and other relevant federal research agencies should assist USPTO in identifying experts to participate in these consultations.**

USPTO is to be commended for the development of its Customer Partnership Program for biotechnological patent applications. The committee urges USPTO to expand the use of input from the scientific community to improve the understanding of the office and its examiners of complex and rapidly evolving technologies, such as genomics and proteomics, with both human health and agricultural applications. The proposed committee should follow the Federal Advisory Committee Act requirements for open meetings and advance notification of meetings.

Nonobviousness

As described in Chapter 3, the *In re Bell* decision is illustrative of the application to genomics of the requirements for nonobviousness. In that case, USPTO argued that a defined gene sequence was obvious from prior art, including the sequence of the encoded protein and a general method of cloning. The inventor argued that the prior art relied upon by USPTO did not suggest all of the modifications to the cited cloning technique that would make it operative and that USPTO had, without supporting evidence, deemed such modifications within the ordinary skill of the field.

In *Bell* and then *In re Deuel* the court held that—as of the time the invention was made—a gene is just another type of chemical compound, and the issue for nonobviousness is the structure (that is, sequence) of the gene. Unless the sequence is predictable from the prior art, the gene is nonobvious. In these two cases, the court refused to see that there is a known relationship between a gene and the protein it encodes.

The National Academies' 2004 report, *A Patent System for the 21st Century,* observed that advances in proteomics have shown that the relationship between DNA sequence and protein *sequence* is predictable, but the relationship to the *structure* of the protein is not. The report noted that newly disclosed protein structures might satisfy the nonobviousness standard more easily than newly disclosed DNA molecules, given that the fine details of the three-dimensional structures cannot be deduced accurately from either the protein or DNA sequence. On the other hand, as more proteomic information becomes publicly available through large-scale projects, the ability to predict the structure based on the amino acid sequence of a protein and the ease with which the structure is obtained will dramatically improve. Nonobviousness determinations require that one look to the prior art and assess whether a person of ordinary skills could replicate the invention, whether such a person would be motivated to do so, and whether he or she would have a reasonable chance of success.

The previous National Academies' committee recommended that the Federal Circuit abandon the rule announced in *Bell* and *Deuel* that, essentially, prevents the consideration of the technical difficulty faced in obtaining pre-existing genetic sequences. The National Academies sought an approach similar to that of other industrialized countries when examining the obviousness of gene-sequence-related inventions: Each case should be analyzed at least in part by looking at the technical difficulty a skilled artisan would have faced at the time the invention was discovered.

Recommendation 8:

In determining nonobviousness in the context of genomic and proteomic inventions, USPTO and the courts should avoid rules of nonobviousness that base allowances on the absence of structurally similar molecules and instead should evaluate obviousness by considering whether the

prior art indicates that a scientist of ordinary skill would have been motivated to make the invention with a reasonable expectation of success at the time the invention was made.

NIH should partner with other organizations (e.g., the Federal Judicial Center) to develop venues for educating judges about advances and new developments in the areas of genomics and proteomics.

Utility Standard

The Supreme Court articulated a strict utility standard in its 1966 decision in *Brenner* v. *Manson*, requiring that a patent applicant show that the invention has "specific benefit in currently available form." The court justified this strict approach by noting that "a patent is not a hunting license. It is not a reward for the search, but compensation for its successful conclusion." But the standard has not been applied in a consistent fashion. Some believe more recent decisions of the Federal Circuit have been less strict about the utility requirement, particularly as applied to biopharmaceutical inventions.

The 2002 report on a trilateral comparative study by the EPO, the JPO, and USPTO (2002 trilateral report) considers the patentability of claims related to the three-dimensional structure of proteins under the laws administered by each of those offices. Each of the three concluded that hypothetical claims to computer models of proteins generated with atomic coordinates, data arrays comprising the atomic coordinates of proteins, computer-readable storage medium encoded with the atomic coordinates, and databases encoded with candidate compounds that had been electronically screened against the atomic coordinates of proteins were not patent eligible. The analysis by USPTO emphasized that each of these hypothetical claims was "nonfunctional descriptive material" and therefore "an abstract idea."

Understanding how genetic variation leads to individual variation in humans is one of the great scientific challenges of the twenty-first century. The path forward will inevitably involve an increasingly broad survey of genetic variation across the genome and establishing the causal relationship of certain regions and ultimately genes with particular traits. Indeed, technology already is in development that would allow complete cataloging of an individual's genetic code at affordable costs. As these technologies are implemented, diagnostics will move from a focus on single genes to a search of all genes.

If it is determined to be essential to allowing research to proceed and medical practice to advance in the coming years, those who are discovering associations between sequence variants and traits should eschew patents. Failing that, the best practices established by NIH and the broader scientific community should be followed. USPTO should require high standards for utility as mandated by existing Supreme Court precedent.

Although the views of USPTO and its foreign counterparts are of enormous practical importance in determining what receives a patent, neither the USPTO guidelines nor the 2002 trilateral report has the status of binding legal authority. As discussed in Chapter 3, the utility standard has proven difficult to administer in a consistent fashion. The committee believes this problem should be addressed.

The committee endorses the USPTO utility and written description guidelines and commends the office for adopting them. The committee also commends the process of input from the scientific community that led to their adoption and modification. Ongoing dialogues of this sort, and as recommended above, should form the basis for continually adapting the guidelines as the underlying science moves forward. However, the scientific community also must bear some responsibility for interpreting the guidelines.

Recommendation 9:
Principal Investigators and their institutions contemplating intellectual property protection should be familiar with the USPTO utility guidelines and should avoid seeking patents on hypothetical proteins, random single nucleotide polymorphisms and haplotypes, and proteins that have only research, as opposed to therapeutic, diagnostic, or preventive, functions.

A move toward a higher standard by the scientific community, USPTO, and the courts would be consistent with the 2001 USPTO guidelines initially adopted to limit patenting of ESTs. Those guidelines recently have been upheld by the Court of Appeals for the Federal Circuit (*In re Fisher).* The committee believes that such guidelines have had a beneficial effect and USPTO should ensure that they are applied to proteomic inventions.

FACILITATE RESEARCH ACCESS TO PATENTED INVENTIONS THROUGH LICENSING AND SHIELDING FROM LIABILITY FOR INFRINGEMENT

Experimental Use Exemption

Academic scientists commonly assume that their research is shielded by law from intellectual property infringement liability (NRC, 1997). However, in *Madey* v. *Duke University*, the Federal Circuit rejected the experimental use defense in the context of academic research, declaring the noncommercial character of the research to be irrelevant to its analysis of the case. The court found that research that is part of the "legitimate business" of the university is not protected "regardless of commercial implications" or lack thereof.[5] The implications of this deci-

[5]*Madey* v. *Duke University*, 307 F.3d 1351 (Fed.Cir. 2002).

sion are not yet clear, although it would appear that researchers and their institutions will have to pay much closer attention to the intellectual property issues involved in their current and future work especially when that work is driven by commercial considerations. Given the nature of much university research—that is, investigator initiated, highly decentralized, and uncoordinated—the implementation of an administrative structure that would deal prospectively with intellectual property issues in a manner similar to due diligence precautions in the private sector could impose burdensome administrative costs and strongly influence choices of academic research directions. At the same time, it is doubtful that such an apparatus could be effective in a university context. The ongoing "research exception" litigation is indicative that many aspects of the law governing patent rights to research tools are not settled.

The committee believes that there should be a statutory exemption from infringement for experimentation *on* a patented invention.

Recommendation 10:
Congress should consider exempting research "on" inventions from patent infringement liability. The exemption should state that making or using a patented invention should not be considered infringement if done to discern or to discover:

a. the validity of the patent and scope of afforded protection;
b. the features, properties, or inherent characteristics or advantages of the invention;
c. novel methods of making or using the patented invention; or
d. novel alternatives, improvements, or substitutes.

Further making or using the invention in activities incidental to preparation for commercialization of noninfringing alternatives also should be considered noninfringing. Nevertheless, a statutory research exemption should be limited to these circumstances and not be unbounded. In particular, it should not extend to unauthorized use of research tools for their intended purpose, in other words, to research "with" patented inventions. Accordingly, our recommendation would not address the circumstances of the *Madey* case, which clearly entailed research "with" the patented laser; but it would shield some types of biomedical research involving patented subject matter.

Patent Pooling

A patent pool is an agreement between two or more patent owners to license one or more of their patents to one another or third parties.[6] A 2000 white paper issued by USPTO promoted their use, stating:

---

[6]See Klein, *supra at http://www.usdoj.gov/atr/public/speeches/1123.html.*

> The use of patent pools in the biotechnology field could serve the interests of the public and private industry, a win-win situation. The public would be served by having ready access with streamlined licensing conditions to a greater amount of proprietary subject matter. Patent holders would be served by greater access to licenses of proprietary subject matter of other patent holders, the generation of affordable pre-packaged patent "stacks" that could be easily licensed, and an additional revenue source for inventions that might not otherwise be developed. The end result is that patent pools, especially in the biotechnology area, can provide for greater innovation, parallel research and development, removal of patent bottlenecks, and faster product development (USPTO, 2000, p. 11).

The committee agrees that patent pooling is an approach that might address some issues of access to patented upstream technology and its possible applications to biomedical research and development and that it should be studied.

Recommendation 11:
NIH should undertake a study of potential university, government, and industry arrangements for the pooling and cross-licensing of genomic and proteomic patents, as well as research tools.

Such proposed sharing arrangements are being pursued in agricultural biotechnology by the Public Intellectual Property Resource for Agriculture and the Biological Innovation for Open Society initiative in different ways. One issue that may be important in the lucrative health field is the willingness of academic scientists to have their inventions pooled if that would reduce or threaten their receipt of the share of royalties as typically are provided by universities.

Ensuring the Public's Health

Although the committee was unable to find any evidence of systematic failure of the licensing system, a few cases of restrictive or refusals to license practices by some companies have generated controversy and disapproval because of the potential adverse effects on public health. Through the Agreement on Trade-Related Aspects of Intellectual Property Rights (TRIPS Agreement), some other countries, such as Belgium and Canada, retain the right to issue compulsory licenses if there is a public health imperative. In the United States, courts have used their equitable powers to deny injunctive relief in cases where health and safety are in issue.[7]

Although this option is rarely used and difficult to implement, the threat that a court might decline to enforce a patent by enjoining its infringement may be

[7]See, e.g., *Roche Products, Inc. v. Bolar Pharmaceuticals Co.*, 733 F.2d 858, 866 (Fed. Cir. 1984); *Vitamin Technologists, Inc. v. Wisc. Alumni Res. Found.*, 146 F.2d 941, 956 (9th Cir. 1945); *City of Milwaukee v. Activated Sludge, Inc.*, 69 F.2d 577, 593 (7th Cir. 1934).

enough to spur patent holders to license on reasonable terms (OECD, 2002). It always should be a last resort, when all else fails, and when protection of the public health cannot be achieved by any other means.

Recommendation 12:
Courts should continue to decline to enjoin patent infringement in those extraordinary situations in which the restricted availability of genomic or proteomic inventions threatens the public health or sound medical practice. Recognition that there is no absolute right to injunctive relief is consistent with U.S. law and with the Agreement in Trade-Related Aspects of Intellectual Property Rights (the TRIPS Agreement).

Gene-Based Diagnostic Testing

Absent special circumstances, such as when the costs of development are high, the licensing of genomic and proteomic tools should be broad so that they ensure patient access and the opportunity to improve upon the method. The committee recognizes that diagnostic tests will sometimes involve such special circumstances and that there is a need to license more exclusively when the costs of test development or diffusion require the substantial investment of private capital. It is likely that with continued advancements in human genomics and the recognition of ever more statistical correlations between mutations in multiple genes and clinical phenotypes, opportunities for engaging in such restrictive practices will continue to multiply. Nevertheless, licenses on genomic- or proteomic-based diagnostic tests, when inventing around the test is not possible, should create reasonable access for patients, allow competitive perfection of the test, not interfere with noncommercial applications of the test in Institutional Review Board (IRB)-approved clinical research, and ensure compliance with regulatory requirements such as permitting quality verification. To ensure a reasonable return on investment, the license may require that the licensee first be given the opportunity to furnish the materials or services required.

The committee recognizes that exclusivity is commonly required to secure the large amounts of investment capital that are needed to establish testing capability on an industrial scale. On the other hand, the exclusive practice of any medical procedure or clinical diagnostic test is an important issue for the medical profession and raises important questions of public health and science policy. For example, the performance of a gene-based clinical test in an academic setting often generates rich databases of newly detected genetic variations that can be correlated with an array of clinical phenotypes. Such admixed medical practice and research provides important new information about the mutational repertory of specific disease-linked genes, as well as the phenotypic correlations that provide new insights into disease mechanisms and identify potential new targets for therapeutic intervention. In instances of the exclusive patenting or licensing of a test, such correlations will only occur if the data derived from the test are made freely available to the clinicians treating the patients. Thus, clinical research in the United States always has been intertwined with the practice of medicine by physician investigators in academic medical institutions, and historically, overages obtained from medical practice have been a significant source for investment and operating funds in clinical research.

Furthermore, the practice of gene-based diagnostic tests by academic laboratories on the large and heterogeneous patient populations of the academic medical center generates rich databases of newly detected genetic variations that can be correlated with an array of clinical phenotypes. Such admixed medical practice and research provides important new information about the mutational repertory of specific disease-linked genes, as well as the phenotypic correlations that provide new insights into disease mechanisms and identify potential new targets for therapeutic intervention. Such research is a hallmark of academic medical practice and historically has made enormous contributions to the advancement of medical knowledge and public health.

It also is the case that health professionals, the biopharmaceutical industries, and the public are anticipating eagerly a new era of "individualized medicine" and the application of pharmacogenomics to guide the drug development process and tailor therapeutic interventions to individuals and populations based on known genetic factors predictive of drug efficacy and safety. For industry to exploit this promising potentiality, the development and practice of precise, gene-based diagnostic tests to identify the candidate populations for both drug testing and marketing will be required. The development of new genetic tests will be linked intimately as never before to drug development, testing, and marketing.

Given the rapid development of gene-based diagnostic testing and its increasingly critical role in the practice of medicine, the committee identified a variety of concerns that it believes should be considered in licensing practices on genomic- or proteomic-based diagnostic tests, where inventing around the test may not be possible, including:

- access for patients;
- allowing competitive perfection of the tests;
- facilitating IRB-approved clinical research in academic medical centers regardless of funding sources;
- facilitating professional education and training;
- permitting independent validation of test results; and
- ensuring regulatory compliance.

Although the committee discussed all of the above concerns at length, it was especially concerned with independent validation of genomic- or proteomic-based test results. Certain members of the medical and academic community noted that, where patent owners may control access to genomic- or proteomic-based diagnostic tests, the patent owners may not allow others to use the patented technologies to validate the results of particular clinical tests. The committee agreed that this may present a problem and encourages patent owners to consider entering into licenses that will permit others to use the patented technologies for the purpose of independently confirming the results of a diagnostic test.

Recommendation 13:

Owners of patents that control access to genomic- or proteomic-based diagnostic tests should establish procedures that provide for independent verification of test results. Congress should consider whether it is in the interest of the public's health to create an exemption to patent infringement liability to deal with situations where patent owners decline to allow independent verification of their tests.

1

Introduction

The advent of the molecular era in biology in the 1940s and 1950s, and in particular the development of the tools of recombinant DNA in the mid-1970s, made it possible for scientists to isolate individual genes and determine their chemical composition and ultimately to sequence entire genomes. The ability to map and sequence genes has not only advanced our fundamental understanding of how genes are assembled into genomes, it also has yielded highly detailed knowledge of the structure of evolutionary trees, increased our understanding of genetics, and led to the development of new diagnostics and therapeutics for diseases such as hypertension and cancer. In recent years, research has progressed beyond creating an inventory of human genes (mapping and sequencing) to efforts aimed at elucidating gene functions, comparing the human genome with those of other species, studying the interactions between genes and the environment, analyzing the structures and functions of proteins encoded by genes, and ultimately determining the role of genes and proteins in human as well as in animal and plant biology.

The sequence of the human genome, which was nearly completed in 2003, is arguably the most powerful dataset the biomedical research community has ever known. Yet its full meaning is just beginning to be revealed. Although human beings may each possess 20,000 to 25,000 genes—far fewer than originally imagined—these genes encode millions of proteins that are responsible for their distinctiveness and that of their families.[1] The challenge for the future is to under-

[1]The "millions of proteins" reflect both a small multiple of the genes via splice isoforms and then a large multiple for post-translational modifications.

stand the information in the genome and to use it to benefit human health and well-being.

One important factor in the realization of the potential benefits of the Human Genome Project (HGP) that requires careful scrutiny is the practice of protecting intellectual property in the fields of genomics and its offspring—proteomics—the study of the protein products of the genome. Patents are sought not only by private sector scientists but also by scientists in universities, research institutes, and government laboratories. Whether the patent claims a gene sequence, its protein product, or a method to detect, produce, study, or manipulate the gene or protein, the freedom of others to conduct research on the role and function of a given gene or protein and their ability to employ them in health care on a reasonable basis could be constrained by the prior existence of a patent, or, more likely, an exclusive license or other restrictive license on a patent.

At the same time, intellectual property protection is essential to biotechnology and pharmaceutical firms that must invest hundreds of millions of dollars in research and development over many years to bring their products to market. To enable firms to garner the sustained investments needed for diagnostic and drug development and testing, patents provide a period of exclusivity with respect to the manufacture, use, or sale of the product. Furthermore, many biotechnology firms have established a market niche between the fundamental research of academic and government laboratories and the more applied research and development activities of large pharmaceutical firms. To remain viable, these companies also rely on intellectual property rights to discoveries that often are made early in the research and development process (i.e., closer to the basic research end of the spectrum) (Eisenberg, 1997). The scale of the rush to protect the rights to new genes is reflected in the fact that by 2001, before the HGP was even completed, just two biotechnology companies alone had filed more than 25,000 DNA-based patent applications for both full-length genes and gene fragments (Service, 2001).

Research universities, too, were spurred by federal legislation enacted in the 1980s to promote the commercial application of fundamental discoveries by their faculties by protecting intellectual property that could be licensed to companies. In a few well-publicized instances, this practice has reaped substantial financial rewards for the universities and inventors, which in turn has motivated other universities to adopt aggressive technology transfer practices. Today, as a consequence of all these activities, some fear that the public good derived from providing incentives to inventors so that they can benefit from their discoveries and from ensuring that public investments in basic research lead to effective prevention and treatment of disease is at risk of being diminished significantly by the negative potential of "thickets" of patents inhibiting future scientific discovery and development.

THE PUSH TO PATENT

The race to patent genes and their protein products in the life sciences began in the late 1970s, with the cloning of genes, the products of which had the potential to be therapeutic products themselves. In that sense, the early patenting of genes encoding proteins such as human insulin, growth hormone, and factor VIII was analogous to the patenting of chemical drugs. To distinguish DNA from a naturally occurring product, the claims of the DNA patents specified recombinant materials, the processes for producing the protein in bacterial or yeast cells, and the material in a form that was "purified and isolated."

In the early 1980s, a series of judicial and administrative decisions clarified patent law, although the statute describing patentable subject matter did not change. In *Diamond* v. *Chakrabarty*, the Supreme Court by a 5-to-4 vote confirmed:

1. that Congress intended patentable subject matter to "include anything under the sun that is made by man" (here the Court quoted from the legislative history of the 1952 Patent Act, the current basic patent law);
2. that "the laws of nature, physical phenomena, and abstract ideas have been held not patentable"; and
3. that "the patentee has produced a new bacterium with markedly different characteristics from any found in nature and one having the potential for significant utility. His discovery is not nature's handiwork, but his own; accordingly it is patentable subject matter under §101 [of the Patent Act]."

Thus, the Supreme Court ruled that a living, genetically altered organism may qualify for patent protection as a new manufacture or composition of matter. In fact, in spelling out its 1980 decision in *Chakrabarty*, the U.S. Supreme Court used much the same definition of patentable subject matter that had been in place since Thomas Jefferson wrote the Patent Act of 1793.

The United States and many other countries already allowed the patenting of products of nature in an isolated and purified state, when their purification led to a new use for that material. Domestic and international patent policies treated DNA sequences as "compositions of matter," much like any other chemical formulae. Thus, the areas of biological discovery emerging from the HGP and related efforts—from gene sequences to proteins—are potentially patentable subject matter as long as the invention meets the standard criteria of novelty, utility, and nonobviousness, describes the invention ("written description"), and provides sufficient detail to enable others "skilled in the art" to make and use the invention ("enabling disclosure"). The "enabling disclosure" requirement mandates the creation of an instructional map that a practitioner in the inventor's field can follow to create and use the invention.

The *Chakrabarty* decision, coming as it did at a time when the cloning and sequencing of genes was becoming increasingly accessible to molecular biology laboratories, further encouraged the patenting of genes and their protein products that were not likely to be therapeutic agents themselves but that could be useful in the development of drugs, research tools, and even genetically altered animals (see Box A). Typically, patents on such basic technology had been considered "upstream" inventions, meaning that a commercial product might not be immediately anticipated and that much further refinement and investment had to occur to reach that point. Such upstream inventions fell into many classes; for example, membrane receptors that could be used to identify agonists and antagonists, proteins involved in biochemical pathways implicated in a disease, and potential extracellular ligands with homology to proteins of known function. Awarding patents for these inventions may offer the possibility for the inventor to participate in any financial benefit that might result from the use of his or her discovery in the development of a drug or other useful product. On the other hand, such upstream patents could be broadly enabling in many different areas of basic research, and, if kept as a trade secret by a single company or exclusively licensed to one or very few companies, they could stymie scientists more broadly in their pursuit of basic knowledge. Patenting these upstream inventions has the advantage, therefore, of assuring universal access if licensed broadly. However, given the unique nature of human genes and the crystalline structures of human proteins, scientists may find it difficult or impossible to "invent around" the subject matter if patented and if the patent can be enforced (i.e., to develop a substitute that allows them to continue working in the art without infringing the patent). If nature provides only one code or structure for a gene or protein, and someone is granted a patent on the discovery and the description of that code or architecture, then other scientists are, at the same time, given access to the new science, and possibly prevented from making certain uses of the patented invention in research.

One class of patents that affects the field of genomics and proteomics describes laboratory methods or procedures and is generally referred to as *process patents*. Prominent examples include the now-expired Boyer-Cohen patents on the techniques of recombinant DNA, held by the University of California, San Francisco, and Stanford University, and the Axel-Wigler patent on introducing DNA into eukaryotic cells, held by Columbia University (both Stanford University and Columbia University allowed nonprofits to practice their patented technology without licenses). These universities licensed the use of the technology widely and nonexclusively to private companies for relatively modest fees, and freely to universities and nonprofit research organizations; thus, the existence of the patents is not believed to have impeded research materially.[2] However, not

---

[2]Columbia University has a new Axel-Wigler patent that it is seeking to exploit and that is the subject of a lawsuit with companies that have refused to take a license for it.

BOX A Patenting the Oncomouse

After the *Chakrabarty* ruling, several critics insisted that the decision appeared to leave no legal obstacle to the patenting of higher forms of life—plants, animals, and possibly human beings—or, by implication, to the genetic engineering of such life forms. Harvard University and Philip Leder moved to take advantage of the legal opening presented by *Chakrabarty*. A distinguished biomedical scientist, Leder was appointed in 1981 to the faculty of the Harvard University Medical School. In conjunction with his recruitment, the DuPont Corporation gave Harvard $6 million for support of Leder's research. The principal *quid pro quo* was simple: Although Harvard would own any patents that might arise from Leder's research, DuPont would be entitled to an exclusive license on any and all such patents.

Over the next two years, Leder and his collaborator Tim Stewart developed a so-called oncomouse—a mouse genetically engineered to be highly susceptible to certain types of cancer. They accomplished this feat by exploiting the then-recently developed transgenic technology to insert the *myc* oncogene, tied to a mammary-specific promoter, into the new embryo of a normal mouse. Leder wondered whether his mice might be eligible for patent protection because they formed a man-made model system for the study of cancer, including the testing of its causes and therapies. Given the *Chakrabarty* decision, Harvard's lawyers saw no legal basis for excluding claims on animals, and on June 22, 1984, on behalf of Harvard University, filed an application for a patent on Leder and Stewart's invention. The main utilities claimed were straightforward, including the use of such animals as sources of malignant or proto-malignant tissue for cell culture and as living systems on which to test compounds for carcinogenicity or—in the case of substances such as Vitamin E—for the ability to prevent cancer. The claims extended to any transgenic mammal, excluding human beings, containing in all its cells an activated oncogene that had been introduced into it, or an ancestor, at an embryonic stage. In April 1988, a U.S. patent was awarded to Harvard University on any nonhuman mammal transgenically engineered to incorporate in its genome an oncogene tied to a specific promoter. It was the first patent on a living animal in the history of the world's patent systems.

all method-of-use patents have been handled in this manner. For example, in the United States, Myriad Genetics holds a patent for diagnostic testing for breast cancer susceptibility based on the BRCA genes. Myriad chose to exercise its patent rights by remaining the sole provider of the test, which indicates whether a person carries a mutation in BRCA genes. In 1997, cancer genetics laboratories—many of which also are research and teaching laboratories linked to major cancer treatment centers—were told to cease providing the tests, which infringed upon the patent. As a result, patients diagnosed by Myriad as positive for one of the two known BRCA genes find it difficult, if not impossible, to turn elsewhere for independent verification of the test results.

SCIENCE AND COMMERCE

Many research scientists who work in public institutions are troubled by the concept of intellectual property protection for DNA-based information, because it seems to be in conflict with scientific norms that dictate that science will advance more rapidly if researchers enjoy free access to knowledge. However, use of the patent system means that there will be less of an incentive to resort to protecting knowledge by making it a trade secret. Patenting entails making public a complete description and a full enabling disclosure of the new technology. The law of intellectual property rests on the assumption that exclusive rights create the ability to attract the investments to fund the research and development required to bring a novel product to market.

The federal government, which supports the vast majority of fundamental biomedical research in the United States, has adopted policies over the past 25 years that are intended to promote the commercialization of research conducted with federal funding as a means to speed the development of benefits to the public good. The Stevenson-Wydler Technology Innovation Act (P.L. 96-480) enables the National Institutes of Health (NIH) and other federal agencies to enter into license agreements with commercial entities that promote the development of technologies developed by government scientists. The act also provides a financial return to the public in the form of royalty payments and related fees. The Patent and Trademark Amendments of 1980 (P.L. 96-517, also known as the Bayh-Dole Act) cede to universities and small businesses the right to claim intellectual property protection for discoveries that result from federally funded research and permit universities and the faculty inventors to derive financial benefit from licensing and royalty payments. Partly as a result of these statutes, a large share of issued DNA-based patents is held by the U.S. government and by universities (Pressman et al., 2005; Michelsohn, 2004).

It is important in addressing scientists' concerns about access to information to recognize not only that patent exclusivity is limited in duration but also that it provides a means of protecting inventions *without* secrecy. A patent grants the right to exclude others from making, using, and selling the invention for a limited term, typically 20 years from the application filing date. But to get a patent, an inventor must disclose the invention fully to enable others to improve upon it. All patents are published upon issuance, and as a result of legislation enacted in 1999, a large majority of patent applications in the United States are published after 18 months.[3] The patent system promotes more disclosure than might occur otherwise if secrecy were the only means of excluding competitors. It is less clear how valid this argument is in public sector research, where publication has long been considered the currency of success and professional advancement.

The principal argument for patenting public sector inventions is the fact that typically, post-invention development costs far exceed pre-invention research expenditures, and firms are unable to make this substantial investment without protection from competition. Patents therefore facilitate transfer of technology to the private sector by providing exclusive rights to preserve the profit incentives of innovating firms. Although many observers have raised questions about the effects on the direction of academic research and the behavior of scientists, in the case of DNA-based patent activities, it is the scope of claims that has generated particular concern among some members of the scientific community. The proliferation of broad claims, including many of dubious validity, raises the prospect that current patent and technology policies, combined with rapidly developing science, might lead to a situation in which technology critical to the development of new diagnostics and therapies could be controlled for commercial gain in ways that threaten to impede unduly such development. For some, this trend stands in contrast to a long-standing norm of the life sciences—to ensure the full access to and use of publicly sponsored research results by making them freely available to the public.

This is not the first time that the goals and language of science have the potential to clash with the goals and language of commerce, but the nature of the "property" in dispute (patents on genomic or proteomic inventions) has generated controversy that creates new challenges for reaching the appropriate balance between the two realms.

In recent years, the controversy has shifted from debates about whether patents on genes, gene fragments or sequences, single nucleotide polymorphisms, haplotypes, or proteins are fundamentally inconsistent with the norms of research science—that is, whether patents on such inventions should be allowed at all—to more nuanced questions about what types of research discoveries should be patented and how proprietary research tools should be disseminated to preserve the benefits of intellectual property, while at the same time minimizing interference with the progress of science and the delivery of medical services (NRC, 1997). Concerns also have been raised about gene patenting, for which the goals are to

[3]The United States accepted this international practice as part of the TRIPS [Trade-Related Aspects of Intellectual Property] agreement. However, applicants who declare their intention to seek only U.S. patent protection may by law opt out of automatic 18-month publication. The overall opt-out rate for all patent applications has averaged 10 percent in the last 3 years. The opt-out rate in patent classification 1600 declined from 7.5 percent in FY 2003 to 5.6 percent in FY 2005. SOURCE: USPTO data.

identify new genes, attempt to identify their function through computerized searches of the genomic database, and then seek utility patents covering these genes based on the resulting insight into the gene's potential function.

PATENT ISSUES

In addition to concerns about the openness of science, challenges are being mounted to the validity of the claims made in some patent applications, particularly those that involve proteins and protein fragments and that concern the value of protein structures in function/utility determinations, as well as the value of computational models versus experimentally deduced structures (Berg, 2004; Vinarov, 2003). If the inventor has a full description and an enabling disclosure adequate to support broad claims, licensing issues become more complex, and the possibility for litigation increases. This has the potential to impede research and raise the costs of commercial development.

With regard to utility, in response to substantial pressure from the scientific community, the United States Patent and Trademark Office (USPTO) in 2001 published a set of examiner guidelines that specify that utility should be "credible, specific, and substantial."[4] One question that has been raised regarding obviousness is whether advances in characterization and purification technologies, computation, and instrumentation have rendered routine a discovery process that was formerly laborious and dependent upon human ingenuity. However, patent examiners are not permitted by the law to take into account the manner by which the inventors themselves arrive at an invention in determining patentability. They are permitted only to reference how a person skilled in the technology might have used routine tools in a routine manner to produce a routine result.

In addition, confusion and delays may ensue when the intellectual property rights necessary to arrive at a commercial end product are held by patentees too numerous or heterogeneous to agree on licensing terms—an "anti-commons" (Heller and Eisenberg, 1998). USPTO suggested in a white paper in 2000 that the solution to some gene patent problems might be the use of patent pools.[5] Pooling related patents could reduce the transaction costs of assembling the patent-protected elements of a research platform. Traditionally, this approach has not been used in biomedical research, but collaborative arrangements designed to yield some of the benefits of pooling are being pursued in certain areas of agricultural research (for example, by members of the Public Intellectual Property for Research in Agriculture [PIPRA] initiative) (Atkinson, et al., 2003).

---

[4]The Guidelines are available on the Federal Register Web site at *www.access.gpo.gov/su_docs/aces/fr-cont.html*, or on the USPTO website at *www.uspto.gov*.

[5]See *www.uspto.gov/web/offices/pac/dapp/opla/patpoolcover.html*.

Another way of containing transaction costs is to ensure that only valid patents are issued and come into play. As new technologies have become subject to patenting and applications and issued patents have grown exponentially in recent years, concerns about deteriorating patent quality—the extent to which patents genuinely represent novel, useful, nonobvious inventions that are described adequately—have come to the fore and have led to proposals for expanding the resources of USPTO, tightening the interpretation of the statutory standards, and instituting a more robust system of expedited post-grant challenges within USPTO rather than the courts. These considerations were the subject of the 2004 National Academies' report, *A Patent System for the 21st Century*, (NRC, 2004) and since then have led to some actions. For example, for the past two fiscal years, USPTO's appropriations have been roughly equivalent to its fee receipts, enabling the hiring of more patent examiners. Legislation (H.R. 2795) introduced in the 109th Congress and the subject of hearings in the House and Senate provides for a post-grant opposition procedure, encourages third parties to submit prior art during an application's examination, and reinforces the applicant's duty of candor.

SECRECY VERSUS OPENNESS

At the inception of the HGP, the public co-funders (NIH and the U.S. Department of Energy) emphasized that, in order to reap the maximum benefit from the program, the human DNA sequence data that it develops should be freely available in the public domain. The National Research Council report that set the stage for the HGP in 1988 stated that "... access to all sequences and material generated by these publicly funded projects should and even must be made freely available..." (NRC, 1988). This principle was reinforced in 1988 by the NIH Ad Hoc Program Advisory Committee on Complex Genomes, which stated the following: "Distribution of and free access to the databases (containing the sequence data) must be fully encouraged. Thus, the data must be in the public domain, and the redistribution of the data should remain free of royalties." In 1996 an international group of public and private sector scientists who were engaged in genomic DNA sequencing passed a unanimous resolution—commonly referred to as the "Bermuda rules"—that "all human genomic DNA sequence information, generated by centers funded for large-scale human sequencing, should be freely available and in the public domain in order to encourage research and development and to maximize its benefit to society." Thus the publicly funded HGP established norms of behavior for the genome community that promoted openness. These principles discouraged the patenting of DNA sequences, even though patents on gene sequences guarantee they will be published. Although patent rights themselves do not necessarily prevent the knowledge or information in the patent from being disseminated freely, they can prevent the information from being used freely.

An early and laudable example of a patent holder adopting practices that promoted the dissemination of a critical research tool was the decision by Stanford University and the University of California to make the recombinant DNA technology developed by Cohen and Boyer available free to all university researchers and to corporate researchers for relatively modest fees, rather than licensing the patent exclusively to a single company (Hughes, 2001). This practice has been endorsed by NIH, which issued guidelines for grantees in handling the dissemination of proprietary research tools (NIH, 1998). Although these guidelines are nonbinding, NIH has the option (for example, when the public health is at risk or whenever the policy and objectives of the statute are better promoted by restricting patents) of intervening with an agency declaration of exceptional circumstances, obviating the statutory patent rights provided to recipients of federal research funding.

Although a laudable and apparently successful goal, such openness is—in and of itself—raising some unexpected challenges with regard to intellectual property. Openness can have the unintended consequence of allowing noninventors to exploit the availability of information. This phenomenon is being referred to in some quarters as "parasitic intellectual property claims" (Collins, 2004) and has led to creative licensing mechanisms aimed at ensuring that the data remain publicly available (that is, the data may be used for any purpose as long as access is not obstructed).

THE ILLUSORY EXPERIMENTAL USE EXEMPTION

Adding to the debates about current patenting and licensing strategies in genomics and proteomics is the prevalence in many research institutions of patent infringement resulting from the erroneous assumption that pre-commercial research is shielded from liability for patent or other intellectual property infringement (NRC, 1997). A recent decision by the Court of Appeals for the Federal Circuit (Federal Circuit) in a suit against Duke University has undermined that presumption, finding that research is part of the "legitimate business" of the university and is not protected "regardless of commercial implications" or lack thereof.[6] It would appear that researchers and their institutions now must pay closer attention to the intellectual property issues involved in their current and future work.

This "experimental use exception" litigation indicates that many aspects of the law governing patent rights to biological materials remain unsettled. Although the United States and other countries have unitary patent systems that ostensibly do not discriminate among technologies, in fact accommodations in USPTO practice and in court decisions have arisen from the needs of differing technologies.

[6] *Madey* v. *Duke University*, 307 F.3d 1351 (Fed.Cir. 2002).

For example, in response to concerns raised by the biomedical community regarding patent applications involving Expressed Sequence Tags (ESTs) and other gene fragments, USPTO in 2001 issued examination guidelines clarifying the utility standard and written description requirements. The Court of Appeals for the Federal Circuit recently affirmed a USPTO rejection of patent claims to ESTs, noting that the USPTO utility guidelines comport with the court's own interpretation of the utility requirement. Meanwhile, some important differences remain between the European and Japanese patent offices regarding the standards applied to biological material applications; and these differences, too, are likely to have effects on the conduct and possibly the location of research. The most important of these differences relates to the lower threshold for nonobviousness of sequence-based claims in the United States, compared to the inventive step criteria used in Europe and Japan.

These and other concerns are forcing questions about current practices in the protection of intellectual property. Patents undeniably have led to the stimulation and promotion of the development of new health care products. However, has this development come at the cost of increased out-of-pocket and opportunity costs, delays, and possibly even the obstruction of some research?

CHARGE TO THE COMMITTEE

NIH asked the National Academies (NAS) to study the granting of intellectual property rights and the licensing of discoveries relating to genetics and proteomics and the effects of these practices on research and innovation. Specifically, NIH asked NAS to study and report on:

1. trends in the number and nature of U.S.-issued patents granted for technologies related to genomics and proteomics;
2. the standards that USPTO and other patent offices (specifically in Europe and Japan) are applying in acting on these applications;
3. the effects of patenting genomic and proteomic inventions and/or licensing practices for inventions on research and innovation; and
4. steps that NIH and others might take to ensure the productivity of research and innovation involving genes and proteins.

Under the auspices of the Academies' Science, Technology, and Economic Policy Board and the Committee on Science, Technology, and Law, a study committee was formed to respond to this charge. The study committee was composed of individuals with a broad range of expertise and practical experience, as well as in-depth knowledge of biomedical sciences and the U.S. patent system. Members include basic and clinical researchers, legal scholars and practitioners, economists specializing in conditions of innovation, biotechnology entrepreneurs, managers of pharmaceutical companies, early-stage investors, medical practitioners, specialists in technology licensing, and specialists in the philosophy and ethics of science and medicine.

The committee met 6 times over a 14-month period. During five of these meetings, the committee invited experts to speak about issues under consideration. In addition, the committee held two workshops, one in Washington, D.C., and one in Bellagio, Italy. The committee also sponsored a survey of research scientists (described in Chapter 4) and conducted its own research on the patent landscape and licensing practices in biomedical research.

Several groups have examined the patent system in great depth (e.g., Federal Trade Commission, 2003; NRC, 2004). This current report does not aim to repeat such an analysis but rather focuses on the unique considerations that arise within the context of genomics and proteomics research.

The committee recognizes that there has been no comprehensive analysis of the impact of intellectual property on genomic and proteomic research involving plants and animals. This was not part of NIH's charge to the committee, which was not composed to address it (although it did include an agricultural economist specializing in intellectual property to provide a perspective on that field). The survey conducted for the panel was limited to biomedical researchers, although the patent data presented in Chapter 4 do not distinguish between human and plant and animal-related material because of the lack of consistent discriminating terms in patent claims.

Some grounds exist for hypothesizing that freedom-to-operate issues are more pronounced in agriculture than they are in biomedical research. The field is not nearly as generously funded; prime research targets are much narrower (focusing on a few high-value crops and animal species); patent ownership is much more concentrated in a few public and private hands. The fact that a few cooperative intellectual property management schemes have emerged in agricultural biotechnology—the Public-Sector Intellectual Property Resource for Agriculture (PIPRA)and CAMBIO, Inc., for example—suggests that some obstacles are perceived, at least for public nonprofit researchers working on applications for nonaffluent markets (Wright and Pardey, 2005).[7] The issues addressed by this committee may merit separate study in the agricultural research context.

---

[7]In a survey of 90 plant biology researchers at four public land grant institutions (U.C. Berkeley, U.C. Davis, U.C. Riverside, and the University of Arizona) Zhen and Wright (2005) found that concern with freedom to operate is focused on a lack of easy and quick access to materials held by others. This is similar to the findings from the survey of biomedical researchers reported in Chapter 4.

ORGANIZATION OF THE REPORT

Following this chapter, the committee provides an overview of the science of genomics and proteomics and discusses policy developments in these fields. Chapter 3 addresses the specific intellectual property issues raised by genomics and proteomics and their interpretation by USPTO and the courts. Chapter 4 presents the results of data collection and analysis activities conducted by the committee. The committee's conclusions and recommendations are provided in Chapter 5.

2

Genomics, Proteomics, and the Changing Research Environment

Since 1944, when Avery, MacLeod, and McCarty published experimental evidence suggesting that DNA serves as the repository of genetic information (Avery et al., 1944), our understanding of the organization and biological function of DNA has increased dramatically. Their revolutionary insight led to the elucidation of the so-called genetic code, which underpins the central dogma of molecular biology: DNA makes RNA (specifically messenger or mRNA), which makes proteins. Subsequently, exploitation of tools from physics and chemistry enabled spectacular advances in genetics, leading to the molecular biology revolution in the late 1970s to early 1980s, and ushered in the era of DNA cloning with its powerful new tools to study biology.

The Human Genome Project (HGP) (with its many spin-offs, such as the SNP Consortium,[1] the HapMap Project,[2] and the Protein Structure Initiative),[3] aims to provide a complete working knowledge of the human genome and, in the longer term, proteomics, which together will provide information and the tools necessary for advancing our understanding of human health and disease. Most recently, the advent of new technologies permitting the simultaneous study of many thousands of genes, messenger RNAs, single nucleotide polymorphisms (SNPs), proteins, or the products of genes in parallel is producing a flood of information and claims about the role of genes in human disease and behavior.

---

[1]See *snp.cshl.org/.*

[2]See *www.hapmap.org/.*

[3]See *www.nigms.nih.gov/psi.*

This new knowledge is revolutionizing the field of medical diagnostics and could yield a powerful arsenal of therapies that offer the promise of cures instead of just amelioration of symptoms. Precisely because of this potential, the rise of genomics and proteomics has generated numerous policy battles, of which disputes about intellectual property are but one.

This chapter provides background information on the science of genomics and proteomics and their impact on the changing paradigm in genetic or personalized medicine and briefly describes some of the policy debates that have ensued regarding openness and access to genomic and proteomic data as they have affected the conduct of science. Chapter 3 focuses more specifically on intellectual property issues affecting these fields as they have entered the U.S. patent system and the courts.

THE IMPORTANCE OF DNA SEQUENCE

After 1953, when Watson and Crick proposed the essentially correct model for the three-dimensional structure of the DNA double-stranded helix (Watson and Crick, 1953), it soon became evident that genetic information stored in DNA was both finite and discrete (or digital) in nature. Knowledge of the order of the four bases—adenine, guanine, cytosine, and thymine (A, G, C, and T)—within each DNA strand, or sequence, of an organism provides full knowledge of all the genetic information passed from one generation to the next. According to Crick, he and Watson speculated about determining the full sequence of human DNA early on but discarded the idea as one that would not reach fruition for centuries (Crick, 2004).

Astounding progress over the ensuing three decades in the discipline now known as molecular genetics, however, proved their pessimistic estimates incorrect. A DNA fragment from any organism can be inserted (or cloned) into the bacterium *E. coli,* which in turn can generate for further study huge numbers of copies of the desired gene fragment. In 1977, the Nobel laureate chemist Frederick Sanger developed efficient methods for using these amplified samples of genetic fragments to determine the sequence of the DNA bases and published the entire sequence of some small viral genomes (Sanger et al., 1977). By the mid-1980s, much of the molecular genetics research community was engaged in isolating and sequencing from particular organisms DNA for individual genes of interest.

Open, facile access to this relatively limited amount of DNA sequence information became an important priority for molecular biologists and molecular geneticists alike. As a result, in 1979 GenBank was established as a nucleic acid sequence database at the Los Alamos National Laboratory and was funded by the National Institute of General Medical Sciences three years later. In 1988, the National Center for Biotechnology Information (NCBI) of the National Institutes of Health (NIH) was organized, and it took over the management of GenBank.

The GenBank database is designed to provide and encourage access to the most up-to-date and comprehensive DNA sequence information to members of the scientific community. Because protein primary structures now are determined mostly by complementary DNA (cDNA) sequence analysis, links between the nucleotide and protein sequence databases are common. GenBank belongs to an international collaboration of sequence databases, which also includes the European Molecular Biological Laboratory and the DNA Data Bank of Japan. Protein sequences are archived in another international consortium, Universal Protein Resource (UNIPROT),[4] which is a central repository of protein sequence and function.

NCBI places no restrictions on the use or distribution of the GenBank data. However, some submitters may claim patent, copyright, or other intellectual property rights in all or a portion of the data they have submitted. There were 37,893,844,733 bases in 32,549,400 sequence records as of February 2004.

THE HUMAN GENOME PROJECT

In an effort to marshal these rapid advances, Robert L. Sinsheimer of the University of California, Santa Cruz, formally proposed in 1985 the possibility of a concerted effort to sequence the human genome. In 1986, Renato Dulbecco, a Nobel laureate and a member of the Salk Institute, made in the pages of *Science* magazine a similar proposal to provide the underpinning for the study of cancer (Dulbecco, 1986). Influential and widely circulated reports by the U.S. Department of Energy (DOE), the congressional Office of Technology Assessment (U.S. Congress, 1988), and the National Research Council (NRC, 1988) all followed and recommended such a project. The NRC report recommended that the U.S. government financially support a project and presented an outline for a multistep research plan to accomplish the goal over 15 years. Soon thereafter, NIH and DOE signed a Memorandum of Understanding to "provide for the formal coordination" of their activities "to map and sequence the human genome." In fiscal year 1988, Congress formally launched the Human Genome Project (HGP) by appropriating funds to both DOE and NIH for that specific purpose.

As envisioned in the NRC report, the HGP did not begin immediately with human sequencing. Instead, the program sought to build infrastructure through a variety of projects. These efforts included the exploration of alternative sequencing technologies, the adaptation of existing technologies to the simpler problem of sequencing smaller genomes of laboratory organisms, and the development of low-resolution maps of the human genome. Other countries—in particular Britain, France, and Japan—also initiated the HGP, and indeed several early successes came from outside the United States.

Despite broad governmental support, the HGP generated considerable con-

[4]See *www.uniprot.org*.

troversy in the scientific community. The shift from traditional, hypothesis-driven, small-laboratory, one-gene-, one-protein-at-a-time science to this new data-driven, large-scale engineering program initially engendered resistance in the molecular genetics community. Even the project's supporters were far from united in their vision of how best to proceed. Many felt that the project would become feasible only with the discovery of completely novel sequencing methods that would be orders of magnitude faster and cheaper than previous methods. Others, particularly Craig Venter, then an investigator at NIH, argued that for the human genome—when much of the sequence was thought to be without function (so-called junk DNA)—a much more efficient strategy would be to sequence only the protein-coding genes through cDNAs, thereby reducing the amount of required sequence by a factor of 10 or more.

Despite conservative expectations, rapid progress was made on many fronts. A framework human genetic map soon emerged, with far greater resolution than initially anticipated. Circular DNA molecules, or vectors, were devised for carrying ever-larger amounts of DNA into bacteria, thereby facilitating construction of physical maps of whole genomes. Adaptation of conventional DNA sequencing approaches to highly automated machines yielded a dramatic expansion in global DNA sequencing capacity that produced in rapid succession the sequence of the first bacterial genome (*H. influenza*), the first genome of a eukaryote, an organism with a cellular nucleus (baker's yeast or *S. cerevisiae*), and the first genome of a multicellular animal (the roundworm *C. elegans*). Given the pragmatic nature of most scientists, it came as no surprise that the enormous utility of these whole genome sequences across the biological scientific enterprise quickly overcame the objections of remaining skeptics. Novel sequencing methods were not required; instead, the basic Sanger method was almost completely transformed by new machines—developed, for example, by Lloyd Smith, Leroy Hood, and Michael Hunkapiller at the California Institute of Technology—and software by others, such as Phil Green, to deal with the data. The early enthusiasm for sequencing cDNAs or their cousins, expressed sequence tags (ESTs), waned as this information proved to be no substitute for the full genome sequence. However, once a full genome sequence was obtained, both cDNA and EST information proved highly useful in finding genes. EST and cDNA sequencing also provided a rapid means of identifying and characterizing some medically significant genes, opening a path to early intellectual property claims. Venter pursued EST sequencing vigorously, and two companies, Incyte and Human Genome Sciences, devoted extensive resources to capturing these sequences and obtaining patent rights to them (see Chapter 3).

Based on this initial flurry of success, the international HGP began the systematic sequencing of the human genome in 1996 on a pilot scale and in 1999 initiated a full-scale effort. Because many investigators wanted to participate in such a historic project, the pilot phase included laboratories throughout the world. The pilot phase was intended to evaluate the cost and quality of the product, select among the variations in sequencing strategies that were still in play, and determine whether performance and economies of scale warranted reducing the number of participants.

Funded participants met in early 1996 to coordinate their efforts. Among the critical decisions made by the group was the adoption of the "Bermuda Rules" as the basis for data sharing and release (see discussion and Box B). In a subsequent meeting, the group also considered a proposal to switch from a clone-based strategy to a whole-genome shotgun, or fragment-based, approach. The potential value of rapid access to large parts of the genome (and therefore genes) was not disputed, but the proponents could not describe a path from the shotgun data to a high-quality complete sequence. The challenges of assembling sequences of individual DNA fragments and in turn assigning all the pieces to specific chromosomal locations in the correct order and orientation were additional concerns. After vigorous debate, the switch in strategy was rejected.

As the pilot phase drew to a close, the successful groups coalesced around a common strategy and methodology, and a few groups emerged as leaders. Economies of scale also were evident. Most importantly, the pilot phase demonstrated that the strategy was capable of producing high-quality sequences in large contiguous blocks at acceptable costs and that costs were continuing to fall. Funding agencies in the United States and the United Kingdom elected to proceed with a full-scale effort, limiting resource allocations to only a small number of highly successful research teams.

Just as these decisions were being made, Craig Venter and the DNA sequencing instrument manufacturer Applied Biosystems, Inc. (ABI) surprised the genomics community with their announcement of a joint venture to sequence the human genome using a whole-genome shotgun approach, in direct competition with the international effort. Unlike the public project, their data were to be held by a company (Celera, Inc.) and initially released only to paying subscribers. Patents would be sought for genes of interest. The scientists leading both the public and private ventures had strong motives to pursue their own courses, and they justified their plans to their funders. A race was on.

On June 26, 2000, the public and private groups announced jointly at a White House-sponsored event that each had succeeded in producing an initial draft of the human sequence, with simultaneous publications describing their findings appearing in 2001 (Lander et al., 2001; Venter et al., 2001). The international HGP published a full and significantly more accurate human genome sequence in 2004. Genome sequences from species across the evolutionary tree continue to flood the databases today.

HUMAN GENETIC VARIATION

One of the most important uses of the human genome sequence information is its explanation of how DNA sequence variation leads to differences among individuals (phenotypic variation) and guidance on how to apply that information for the betterment of humankind. The development of genetic and physical maps and the ongoing release of the human sequence over the past 15 years have greatly increased the number of genetic diseases for which the causative defective, or mutant, gene has been identified. Today the genetic bases for all the major Mendelian (single gene) diseases are known. OMIM (the Online Mendelian Inheritance in Man) now lists approximately 2,000 genes in which the molecular basis for the Mendelian phenotype is known. Just 15 years ago, only a handful of such genes were known, and the cloning of a gene responsible for human genetic disease became front-page news.

Such molecular insights into disease are leading to new strategies for diagnosis and therapy. Definitive diagnoses can be made directly on the defective genes themselves, without the ambiguities of previous indirect phenotypic measures. DNA testing also can be carried out prospectively, permitting action to be taken before overt symptoms develop, an important advantage in genes that predispose individuals to cancer, for example. Tests can even be conducted prenatally, as early as the pre-implantation stage of development, or even *in vitro*, allowing prospective parents a choice—a significant benefit in cases of devastating childhood genetic diseases such as Tay-Sachs, sickle cell anemia, or cystic fibrosis.

Exploiting molecular insights with which to craft alternative therapies has proven to be more challenging than developing new diagnostic tools, but important progress is being made. The most obvious gene-based strategy is called "gene transfer" or "gene therapy," which involves correcting the underlying genetic defect by providing a patient's cells with a functional gene that directly reverses the deleterious effects of the mutant or missing gene. To date, success has been limited to a small number of relatively special cases. More encouraging, however, is the growing realization that our knowledge of the precise molecular nature of the genetic defect or mutation can lead to specific therapies that block the consequences of the mutation *indirectly*. For example, the discovery and characterization of the chromosomal fusion that causes white blood cells to divide uncontrollably, giving rise to the cancer known as chronic myelogenous leukemia (CML), eventually led to the development of the small-molecule drug imatinib (Gleevec®/Glivec®), which has yielded spectacular results in patients who would otherwise have died within a few years of diagnosis. Imatinib works by blocking the inappropriate function of a fusion protein, BCR-ABL, which is encoded by a new gene that is created by fusion of two chromosomes in the patient's leukemic cells. Alternative treatment strategies are aimed at trying to restore protein functions that have been lost as a consequence of mutations. For example, drugs that directly influence the functioning of mutant forms of the cystic fibrosis transmembrane conductance regulator protein are now going into clinical trials with the hope that they will be able to restore sufficient function to alleviate the devastating symptoms of cystic fibrosis.

We know, however, that human genetic variation is at the root of many more diseases than these relatively rare single gene disorders. Detailed comparisons of human DNA sequences have demonstrated that two copies of the human genome differ by about 1 base in every 1,300. In all, there are approximately 3 billion bases in the human genome, which means that the DNA sequences of any two individuals differ at more than 2,000,000 base positions along the DNA double helix. Among such differences are those that underlie heritable variation among individuals for an enormous number of traits, such as eye and skin color. In addition, combinations of particular genetic variations within populations give rise to genetically complicated or multifactorial diseases (e.g., hypertension, colon cancer). Medical geneticists are just now beginning to use comparative human genome sequencing to understand the extent of genetic variation and to describe the common variants shared across populations. Two prominent extensions of the HGP, the SNP project and the HapMap project, have begun to build this foundation. Indeed, understanding how genetic variation leads to individual human variation is one of the great scientific challenges of the twenty-first century.

The path forward will inevitably involve an increasingly broad survey of genetic variation across the genome in larger and larger groups of individuals. Correlation of genetic and phenotypic differences will establish causal relationships, ultimately revealing the identities of the multitude of genes that contribute to particular traits. Methods for assaying genetic variation are changing rapidly, with various revolutionary approaches nearing commercial testing. The most impressive of such innovations would allow for the complete cataloging of an individual's DNA sequence at a cost of less than $1,000 per person. As these cutting-edge technologies are introduced and an increasing number of causal relationships are known, the field of diagnostics will move from its current focus on single genes to a search of all the genes responsible for a particular disease. This knowledge will be critical to realizing the goals of personalized medicine, among other potential benefits, in which drugs are targeted to small groups and even individuals who are likely to benefit from the therapy and unlikely to suffer adverse reactions. For example, Genzyme has just introduced a test (based on research from the Massachusetts General Hospital and the Dana-Farber Cancer Center) that identifies cancer patients who are more likely to have a favorable response to cancer drugs that target the epidermal growth factor receptor.

The scientific community will need freedom to operate to realize these achievements, but concerns exist about the multitude of existing patents on genes and fragments of the human genome—with the prospect of even more—that could impede or even block progress. Early-stage applications are likely to be affected more severely.

WHAT ARE GENOMICS AND PROTEOMICS?

The success of the initial phase of the HGP and the attendant availability of the human genome sequence and the genomes of numerous other organisms have transformed the study of biology. Most obviously, the full catalog of genes from each genome opens up new avenues of study. No longer does an investigator need to confine his or her inquiries to a single gene or a small set of genes; instead, the behavior of an ensemble of genes can be investigated simultaneously. At the level of science policy, the genome projects have served to validate data-driven or discovery-based approaches as legitimate intellectual competitors of more traditional hypothesis-driven research programs. Finally, at the level of methods development and scientific instrumentation, the genome project made "high-throughput" methods, including robotics and sophisticated computing, part of the biologist's standard toolkit. This constellation of attributes—comprehensiveness, data-driven character, and large scale—distinguishes genomics from its parent science, genetics.

The characteristics of comprehensiveness, scale, and intellectual attitude differentiate proteomics from more traditional ways of studying proteins in much the same way. Interactions between pairs of proteins can be evaluated, not just those of a few likely candidates. Protein identification using mass spectrometric analysis can compare the patterns obtained to what is possible in the genome and quickly identify many of the proteins in a particular mixture, rather than exhaustively characterize each constituent one at a time. In a subdiscipline of proteomics known as structural proteomics, the three-dimensional structures of proteins can be examined in systematic, data-driven projects, rather than focusing on the precise details of a single protein. Indeed, "-omics" has become shorthand for data-driven, large-scale, comprehensive projects of a enormous variety.

The results of genomics and proteomics increasingly promise the potential for future widespread adoption in medicine and biology. Simultaneous measurement of many mRNA levels now can reveal patterns of gene expression for an organism or a tissue under various conditions that can then be compared, pointing to genes characteristic of certain states or reactions. For example, distinct subtypes of large-cell lymphomas, with quite different responses to chemotherapy, can be distinguished from one another by measuring mRNA expression patterns, thereby providing a means of directing therapy (Staudt and Dave, 2005). Likewise, serum samples can be decomposed into a spectrum of proteins, looking for patterns—referred to as biomarkers—of a particular disease, opening up the possibility for early detection and diagnosis.

The DNA found within each cell contains the genetic blueprint for the entire organism. Each gene contains the information necessary to instruct the cellular machinery how to make mRNA, and in turn the protein encoded by the order of bases constituting the gene. Each one of these proteins is responsible for carrying out one or more specified molecular functions within the cell. Differing patterns of gene expression (i.e., different mRNA and protein levels) in different tissues explain differences in both cellular function and appearance.

The original central dogma of molecular biology posited that each gene encodes the information for the synthesis of a single protein. In recent years, however, it has become apparent that the primary RNA transcript of eukaryotic genes can be processed in more than one way—one gene can produce more than one protein. In one dramatic case, it is thought that a single gene in *Drosophila* has the potential to encode more than 30,000 closely related but distinct proteins (Gravely, 2005). Furthermore, once a protein is produced through a process called translation, it can be further modified by the covalent attachment of substances such as sugars, fats, phosphate groups, and other so-called post-translational modifications that affect the function that the protein performs for the cell. Together, these various possibilities constitute the proteome—the entire set of proteins made by a cell or, in the case of multicellular organisms, such as humans, all the cells of the body—that is many, many times larger and much more complex than its corresponding genome.

In addition to addressing the full complement of genes in a genome, genomics involves global studies of gene expression, including expression patterns of particular cell types and gene expression under specific conditions or during particular stages of development. For example, muscle cells preferentially express the gene encoding the red-colored, oxygen-storage protein myoglobin at high levels, thereby ensuring that we can exercise. In contrast, skin cells strongly express the genes encoding the skin keratin proteins that provide the protective layer covering our bodies. Skin cells do not express myoglobin, and muscle cells do not express skin keratins. Such genes are said to be silent or "switched off." Thus, each cell, despite having the same genome, utilizes a different complement of expressed mRNAs, or "transcriptome."

Like genomics, proteomics aims to study the entire repertoire of proteins within an organism. Proteomics is far less advanced than the field of genomics because robust technologies to study the structure and prevalence of all proteins in a cell in a high-throughput manner are only now being fully developed. The challenge that proteomics faces is enormous because of the finding that many genes code for multiple proteins, and those proteins are modified post-translationally in complex ways. As with genomics, proteomics has much to tell us about complex disease states and our own evolution. Differences in protein levels and protein modifications can be measured by two-dimensional gel electrophoresis, mass spectrometry, and protein microarrays. Researchers have claimed that measured differences in proteins from the blood of different patients can be used to predict the onset of ovarian cancer (Petricoin et al., 2002), although these approaches yet must be demonstrated to be reliable. Furthermore, scientists using a variety of experimental approaches can determine which proteins interact with one another while performing their cellular functions.

Today, such tools are being used to understand the protein composition of complex biological networks that are responsible for carrying out complicated tasks within cells. In some cases, these networks of proteins are used to transmit signals from the surface of the cell to the nucleus, where they can switch genes on and off. Proteins making up such signal transduction pathways have emerged as important new targets for cancer drugs because of the causal link between the expression of certain genes (e.g., BCR-ABL in CML) and uncontrolled cell division.

THE IMPORTANCE OF PROTEIN STRUCTURE

Unlike DNA, which has a double helical structure regardless of its sequence composition, proteins assume three-dimensional shapes that depend on their precise sequence of amino acids and the microenvironment of the protein. The three-dimensional shapes give rise to the various functions carried out by proteins—the building blocks, machines, and control networks of organisms. These three-dimensional shapes (protein structure) can be determined, as was first accomplished in 1957 for myoglobin by Sir John Kendrew and co-workers (Kendrew et al., 1958), using a physicist-invented technique called x-ray crystallography. In this method, an x-ray beam is directed through a crystal composed solely of the protein of interest. The spray of x rays emerging from the crystal (the diffraction pattern) can be analyzed using computational methods to provide a full atomic description of the three-dimensional shape of the protein. In the 1980s another method, called nuclear magnetic resonance (NMR), was introduced into use to determine the structures of small proteins. Most recently, cryo-electron microscopy has been used to visualize the three-dimensional shapes of collections of proteins found within cells that are referred to as macromolecular "machines."

An early example of the importance of three-dimensional structure in biology was first appreciated when Watson and Crick showed that DNA strands are usually organized in pairs into a double helix, which immediately suggested how the genetic information stored in DNA can be imparted equally to two different daughter cells upon cell division. Subsequently, Kendrew's structure of myoglobin explained how this protein carries out the specialized task of oxygen storage in muscle cells. Simply put, and in principle, "function follows form" in biology, to recast a phrase borrowed from modern architectural theory.

Insights coming from structural proteomics hold the promise of understanding biological processes at the molecular level. The other important practical advantage of knowing the three-dimensional structure of a protein is that small-molecule drugs can be engineered to fit into regions of the protein that are responsible for cellular activities. By using the protein structure to guide optimization of small-molecule drugs, pharmaceutical and biotechnology companies are seeking to develop drugs that preferentially bind to their target proteins and do not cause unwanted side effects by binding to other proteins in the body.

As the number of protein structures elucidated with x-ray crystallography and NMR increased, a need was perceived by the scientific community to archive the atomic coordinates (or positions) for each protein. In 1971, the Protein Data Bank (PDB), the worldwide repository for three-dimensional biological macromolecular structure data, was established at Brookhaven National Laboratories with just seven structures; each year a handful more was deposited. In the 1980s, the number of deposited structures began to increase dramatically, because of improvements in technology for all aspects of the crystallographic process, the addition of structures determined by NMR methods, and the emergence of data-skewing norms in the relevant scientific community. By the early 1990s, the majority of scientific journals required the deposition of atomic coordinates into the PDB and a PDB accession code as a condition of publication. At least one U.S. funding agency (the National Institute of General Medical Sciences) adopted guidelines published by the International Union of Crystallography requiring data deposition for all three-dimensional protein structures. In 1998, the management of the PDB moved to the Research Collaboratory for Structural Bioinformatics—a consortium involving the University of California San Diego and Rutgers, the State University of New Jersey. In 2005, there were more than 30,000 structures in the PDB, and more than 10,000 researchers in biology, medicine, and computer science access the PDB Web site daily. In 2004, the journal *Molecular & Cellular Proteomics* introduced guidelines for authors planning to submit manuscripts containing large numbers of proteins identified primarily by multidimensional liquid chromatography (LC/LC) coupled online with tandem mass spectrometry (MS/MS), or LC-MS/MS (Carr et al., 2004). The guidelines address the need for the scientific community to make such data readily available.

CHANGING SCIENTIFIC AND CLINICAL PARADIGMS

In the last decade, these advances in genomics and increasingly in proteomics have combined with technical advances in molecular biology, liquid-handling robotics, miniaturization, image analysis, and computing platforms to transform the way in which biologists approach the study of cells and even entire organisms. Together they have produced a relentless movement away from an earlier necessary penchant for reductionism toward the goal of understanding how entire biological processes work and are regulated—the goal of systems biology. Today, systems biologists study the complex interplay of a host of genes as these genes give rise to a disease symptom, such as hypertension, or analyze hundreds of proteins in a blood sample to identify patterns that may be indicative of a particular cancer.

This movement toward a more "holistic" view of biology already has begun to change the face of the academic biological research enterprise. Interdisciplinary research teams involving biologists, chemists, physicists, mathematicians, and computer scientists are coming together in growing numbers. As discoveries stemming from genomics/proteomics are transformed into valuable items of intellectual property owned by universities, the new generation of biologists will wield ever more influence.

Clinical medicine and the pharmaceutical/biotechnology companies industry are faced with yet more disruptive influences from the genomics/proteomics revolution. Within a decade, DNA sequence information will become integral to the practice of medicine. Prescription drug usage ultimately may be dictated by a given patient's genetic makeup. Instead of trying to develop therapeutics to relieve *symptoms*, drug companies will come under increasing pressure to tailor therapies to individual groups of patients sharing a particular genomic/proteomic signature or fingerprint (as well as certain nongenetic traits). This sea change will first become apparent in the design and execution of clinical trials, in which genetic predispositions to therapeutic benefits and risks will be analyzed. The benefit to pharmaceutical manufacturers will be cheaper, faster clinical trials promising a higher likelihood of detecting a positive signal and a reduction in the number of adverse events, leading to more rapid approvals. Such advances will, however, come with a price for pharmaceutical and biotechnology companies. The advent of personalized medicine may well bring an end to the era of so-called blockbuster drugs, because product development will be restricted to smaller target patient populations. To say the least, the current economics of drug discovery, development, and marketing will change considerably.

Until very recently the field of human genetics has been restricted to studying diseases whose etiology can be traced to mutations in a single gene (e.g., cystic fibrosis). However, few common diseases are monogenic. The emerging focus of personalized medicine increasingly is on polygenic disorders and the importance of epigenetic changes in disease or the role of a family of somatic cell mutations and epigenetic changes in tumors. In these polygenic disorders, multi-analyte tests will be required, and the relative contributions of each genetic or epigenetic change will need to be defined. For statistical reasons, relatively large populations will need to be studied to define the relative contributions of each change in the DNA sequence or degree of methylation, mRNA expression level or protein concentration, and degree of post-translational modification. The opportunity to create new intellectual property is rich, and it is possible that it will be difficult for one gene patent to block the development of these tests. For example, in gene expression arrays, many genes tend to show highly correlated expression levels; thus, it is often possible to substitute one gene for another without a substantial loss of predictive power. Because of the need to conduct substantial clinical trials to prove the practical value of these poly-analyte tests, however, the companies that develop these tests will have proprietary products.

INTELLECTUAL PROPERTY AND COMMERCIALIZATION

In the 1980s, two series of events converged with the rapid advances in science to transform the conduct of academic biological science: the development of public policies that encourage—even require—scientists and their institutions to pursue commercialization of research, and the growth of the biotechnology industry, which has benefited immensely from the intellectual capital found in academic institutions and the basic research investment made by the U.S. government in biological and biomedical research. These trends, described briefly below, radically altered the culture of biology and challenged its longstanding norms of sharing and openness.

Technology Transfer Policies

In a message on October 31, 1979, and again in his State of the Union Address on January 21, 1980, President Jimmy Carter urged Congress to spur industrial innovation by enacting a three-part reform in the policy and operation of the patent system. One part of the legislative package urged the creation of a uniform government patent policy for university and small business federal contractors and grantees under which they could retain ownership of patents arising from research performed with federal support. Although the presumption was that title to patents produced by other contractors—for example, large corporations—would be kept by the government, these contractors could be granted an exclusive license for commercial exploitation of the invention.

In 1980, in response to concerns about U.S. competitiveness in the global economy, Congress enacted two laws that encourage government-owned and government-funded research laboratories to pursue commercialization of the results of their research. These laws are known as the Stevenson-Wydler Technology Innovation Act (P.L. 96-480) and the Patent and Trademark Amendments of 1980 (P.L. 96-517), the latter also known as the Bayh-Dole Act. Their stated goal is to promote economic development, enhance U.S. competitiveness, and benefit the public by encouraging the commercialization of technologies that would otherwise not be developed into products because of the lack of incentives associated with exclusive rights.

The Stevenson-Wydler Technology Act, which established basic federal technology transfer policies, enables NIH and other federal agencies to execute license agreements with commercial entities in order to promote the development of technologies discovered by government scientists. The act also provides a financial return to the public in the form of royalty payments and related fees. In 1986, the directives of this act were augmented by its amendment, the Federal Technology Transfer Act of 1986 (FTTA), which authorizes federal agencies to enter into cooperative research and development agreements with nonfederal partners to conduct research. The FTTA also authorized federal agencies to pay a portion of royalty income (currently a maximum of $150,000 per inventor per year from all royalty sources) to inventors who had assigned their rights to the government. These payments are not considered outside income; rather, they are deemed part of the employee's federal compensation.

The Bayh-Dole Act was designed to address barriers to commercial development affecting nongovernmental entities, with the aim of moving federally funded inventions toward commercialization. The act enables grantees and contractors, both for-profit and nonprofit, to choose to retain title to government-funded inventions, and it charges them with the responsibility to use the patent system to promote utilization, commercialization, and public availability of inventions. Other provisions ensure among other things that sponsoring agencies have a non-exclusive license to use the invention for government purposes; that nonprofit organizations cannot assign rights to the invention without the approval of the sponsoring federal agency; and that organizations other than small businesses will be prohibited from granting exclusive rights to the invention from the earlier of 5 years from its first commercial use or 8 years from the date of invention. The law also empowers any federal agency to require inventors or their assignees to grant licenses in order to (1) achieve practical application of the invention in its field of use; (2) alleviate health or safety needs; (3) meet requirements for public use specified by federal regulations; or (4) achieve participation by U.S. industry in the manufacturing of an invention. And the law prohibits licensing that reduces competition.[5]

Recipients of federal research funds—academic institutions and industry—have 25 years of experience in technology transfer under Bayh-Dole. To accomplish technology transfer, institutions typically seek patent protection for inventions arising from their research and license rights to private entities in order to promote commercialization. In this way, private entities interested in practicing an invention in which they have no ownership may by entering into a licensing agreement with the patent owner obtaining rights to use and commercialize it.

Because most universities share a substantial portion of the royalty income generated from patent licenses with faculty inventors, patents offer an additional incentive for researchers to pursue projects that have commercial potential. Although the rules that universities use for allocating royalties vary, a typical payment system gives a first cut from royalty income to the university to reimburse it for the costs of filing the patent. After costs are recovered, the income is then divided among the university's technology transfer office, the faculty members listed as inventors, the faculty members' departments, and other departments in the university. Some of these agreements can provide faculty members with as much as 50 percent of the total royalty revenue after patent costs are recovered.

Patenting also increases incentives for faculty members to keep their findings secret until a fully developed patent application or a provisional application is filed. Secrecy can be problematic for the careers of students and junior faculty members who must publish their research findings to establish their reputations and obtain funding. For this reason, most universities strive to file patent applications quickly so that publications are not delayed. Patents, because they are not validated by other academics, may not be a source of academic credit, even though

[5]P.L. 96-517. Summary at *http://thomas.loc.gov/cgi-bin/bdquery/z?d096:HR06933:@@@L%7CTOM:/bss/d096query.html%7C*, consulted August 26, 2005.

they may be sources of credit in commercial science. U.S. patent law provides a grace period enabling an inventor to disclose an invention—for example, in a paper or conference presentation—up to one year before filing a patent application. Other countries do not have this one-year grace period, so a truly valuable invention with worldwide implications cannot be disclosed before the inventor applies for a patent if foreign patents are to be sought. Many universities have adopted regulations limiting the time that commercial sponsors can delay publication so that patents can be filed. Little, however, can be done to prevent faculty members themselves from delaying disclosure of their research to protect their own interests as potential inventors (Blumenthal et al., 1986; 1996; 1997).

Emergence and Expansion of the Biotechnology Industry

The importance of the university scientist to commercial biotechnology has been well documented (U.S. Congress, 1988; NRC, 1988; Blumenthal et al., 1996). Early concerns about collaborative research arrangements in biotechnology, particularly those involving universities and industry, focused primarily on issues of academic freedom, proprietary information, patent rights, and other potential conflicts of interest among collaborating partners (Blumenthal et al., 1986; Kenney, 1986; Kodish et al., 1996). Biotechnology firms and large pharmaceutical companies, however, continue to support biotechnology research in universities.

Concerns persist regarding the subtle impacts of these collaborative arrangements, specifically whether university-industry relationships adversely affect the academic environment of universities by inhibiting the free exchange of scientific information, undermining interdepartmental cooperation, creating conflict among peers, or delaying or completely impeding publication of research results (Firlik and Lowry, 2000).

Several drawbacks to university involvement with industry-sponsored research have been identified (Dueker, 1997; Firlik and Lowry, 2000). University officers and faculty are concerned about constraints on academic freedom and the inevitable conflict between commercial trade secrecy requirements and traditional academic openness. Many circumstances and forces at play prompt companies to control research conduct and protect the secrecy of research data. In surveys of the biotechnology industry conducted in the 1990s, 56 percent of companies reported that in practice, the university research they supported often or sometimes resulted in information that was kept confidential to protect its proprietary value beyond the time required to file a patent (Blumenthal et al., 1997). On occasion, conflicts between companies and faculty about the content in published reports of industry-sponsored research have been reported. For example, companies have preferred not to publish the results of studies resulting in less-than-optimal data, although academics asserted that the insights to be gained by publication would advance scientific understanding (Bodenheimer, 2000).

BOX B Birth of the Biotechnology Industry

The birth of the modern biotechnology industry can be traced to the early 1970s, with the discovery of genetic engineering techniques, such as recombinant DNA methods and hybridoma production. These discoveries were made by biochemists and molecular biologists, many of whom were working at large academic medical centers.

The formation of Genentech is often considered the starting point of the biotechnology industry. Genentech was founded in 1976 by University of California, San Francisco, scientist Herbert W. Boyer and venture capitalist Robert Swanson. In 1978, the company announced that it had successfully cloned a human insulin gene using recombinant DNA technology. This discovery was licensed to Eli Lilly, the largest U.S. producer of insulin, and in 1982 recombinant human insulin was the first recombinant drug to gain Food and Drug Administration (FDA) approval. Human insulin was considered a significant advance in the treatment of diabetes, since a number of diabetics were allergic to traditional insulin extracted from the pancreatic glands of pigs and cows.

Genentech went on to develop and market its own recombinant drugs, the first being recombinant human growth hormone, which was approved in 1986 for use in children with a rare form of dwarfism caused by a lack of sufficient endogenous growth hormone. Prior to the development of Genentech's human growth hormone, these children were treated with growth hormone obtained from cadaver pituitary glands. Problems with this material included periodic shortages and also, rarely, the development of a lethal neurodegenerative disease called Creutzfeldt-Jakob disease, which came from an undetectable infectious agent found in cadaver pituitary tissue.

Another example of the medical advances and commercial successes that could be obtained from genetic engineering comes from Baxter's recombinant factor VIII (Recombinate), which was developed by the Genetics Institute, then licensed to and manufactured by Baxter. Factor VIII is a blood coagulation protein missing in hemophilia A, the genetically inherited bleeding disorder that afflicts about 20,000 males in the United States. Prior to the availability of recombinant Factor VIII, the protein was collected from pooled human blood which, prior to the use of the HIV test in 1985, was often contaminated with the AIDS virus. As a result, almost all hemophilia A patients who received Factor VIII from pooled human blood before 1985 were infected with HIV, and many have died of AIDS. Similarly, hemophiliacs had also contracted hepatitis when these viruses contaminated the blood pool. Recombinant human Factor VIII, approved in 1992, eliminated the constant problem of blood contamination and offers lifesaving benefits to hemophilia A patients (Kaufman, 1989).

continued

BOX B Continued

Recombinant insulin, growth hormone, and Factor VIII typify the advantages that can be achieved with biotechnology. These products proved to be significantly better than previous medical treatment options, and many of the new biotechnology therapies were the first treatments available for a given disease. The drugs were patentable and could command premium pricing, which helped to offset the high development and manufacturing costs and the relatively small market for the diseases treated with recombinant proteins (Thackray, 1998).

Today genomics companies are often divided into large-scale sequencers, positional cloners, and those that do functional genomics. Large-scale sequencers, such as Human Genome Sciences, Inc., develop research databases of genes, gene fragments, or gene expression patterns, which enable drug discovery. Celera Genomics entered this field as it began its human genome sequencing.

Positional cloning companies study the genomes of individuals from families that have specific diseases and try to determine which genes cause the disease. From this information, disease genes can be identified, and tests to detect them can be developed. Companies such as Myriad Pharmaceuticals and Millennium Pharmaceuticals perform this kind of work.

Functional genomics companies conduct research to identify the function of genes. For example, they compare the genes in humans to those in other species, which is valuable because genes often perform the same function regardless of the species, a phenomenon called homology, and it is usually easier to assess gene function in smaller organisms. "Tool" companies, like Affymetrix, develop "array technologies" that can analyze rapidly which genes are expressed in a given tissue or cell. By comparing differences in gene expression between diseased and healthy tissue, this technology is used to discover genetic changes leading to disease.

Notwithstanding the early successes of Celera, Millennium, and Affymetrix, current business models in the biotechnology industry have shifted dramatically from the halcyon days of the genomics company bubble (1999-2000). Today, it is generally believed that long-term value creation in biotechnology can come only from the sale of pharmaceutical products. Most of the so-called platform companies of the late 1990s have disappeared or migrated to drug discovery. The prescient (not to mention lucky) few, such as Perlagen, have evolved into product companies with varying degrees of reliance on their original platform technologies.

On the other hand, much of this research activity is mutually beneficial and also advantageous to society, because it can accelerate the commercialization of useful medical products. Such collaborations also allow companies to remain up-to-date and informed of discoveries as they come off the bench. For clinical research, such as human drug trials, access to patients at university hospitals can be a primary motivator for collaborative research.

Further, biotechnology advances have frequently returned new science and new products to universities for basic and translational research. The production of recombinant erythropoietin (EPO) by Amgen in the late 1980s provides one such example. EPO is the hormone that regulates red blood cell proliferation. Although it had been discovered several decades before, it had never been available in sufficient quantities for many studies. The availability of the recombinant protein enabled such research.

From the university perspective, working with industry carries real benefits. In the medical and agricultural sciences, universities have benefited from research collaborations with industry, especially in the face of limited government research funding. Increased access to resources allows universities to expand research programs, attract new faculty, build facilities, purchase equipment, and enhance their reputations. Industrial relationships bring valuable equipment and prototypes to the university laboratory. Collaboration with industry also provides faculty with an understanding of industrial problems, enriching the training of engineers and scientists for their future work in an industrial environment.

SCIENTIFIC NORMS AND EVOLVING SCIENCE POLICIES

Since the inception of the HGP, debates about access to data and information have remained ongoing. In addition, there have been disputes about the propriety of patenting genes, partial genes, or gene products, as well as disagreements in the research community about whether academic institutions should be encouraging patenting and licensing strategies versus facilitating open access to data and resources. Early battles about the appropriateness of patenting ESTs have led some in the scientific community to be skeptical about the general direction of patent policy in this area.[6] Moreover, controversy lingers over whether research organizations should be exempt from infringement when using patented products or tools in the course of research (as discussed further in Chapter 3). Collaboration with industry may be encouraged, however, by the recent *Merck KGaA* v. *Integra Lifesciences I, Ltd*[7] case, which protected researchers and their academic institutions from litigation, as well as the company that paid for the work, be-

[6]See, for example, *Nature* 396:499.

[7]2005 U.S. Lexis 4840 (2005).

cause the work was related to its FDA submission.[8] This decision seems suggestive of a growing understanding of the need to draw clearer lines regarding what conduct may, or may not, constitute patent infringement.

As science has developed and commercial activities have increased, scientists from both the public and private sectors have been addressing policy issues and appropriate norms of behavior for genomics-related research. As described below, NIH has responded to these concerns by issuing a number of guidance documents that aim to promote the sharing of both resources and data.

Concerns About Openness and Access

The tradition of sharing materials and results with colleagues speeds scientific progress and symbolizes to the nonscientific world that the goals of science are to expand knowledge and to improve the human condition. One reason for the remarkable success of science is the communal nature of scientific activity. Thus, undue restrictions on data, information, and materials derived from science, especially publicly funded science, has been a theme of many discussions in the science policy community over the past 20 years.

Science builds on previous discoveries, with dissemination of discoveries through publication a crucial part of the process. Publication in journals has its roots in the 17th century, with the initiation of the *Philosophical Transactions of the Royal Society* by Henry Oldenburg. Publication does more than bring together individuals who would otherwise work in isolation from one another; it provides a record of the collective body of scientific knowledge. The expectation is that a publication will contain a detailed description of the methods and materials sufficient enough that others can attempt to replicate the results, and if validated, build upon them without constraint. "Science is fundamentally a cumulative enterprise. Each new discovery plays the role of one more brick in an edifice" (Lander, in NRC, 2003, p. 29). Because publication and the implied unrestricted use of the results of research are central to the activity of science, the norms associated with the dissemination of results have been reinforced both implicitly and directly, with an NRC report serving as a recent example.

Publication shares with patenting a similar purpose—inducing the investigator or inventor to reveal the discovery in order to advance knowledge. Both systems expect the description to be detailed enough to allow replication. But the social contract of the exchange is quite different. Patents give the holder the right to exclude others from making, using, or selling the patented product or process in return for disclosure; thus, patents constitute a limited monopoly. Publication ascribes credit to the authors for the primacy of their discovery with its attendant

[8] *Id.* at 13-14.

benefits in exchange for unconditional use of the discovery, including the materials and methods, for the benefit of science. Publication, of course, does not preclude patenting (or vice versa), and from a social perspective the two systems are complementary: Patenting fosters commercialization of ideas, and scientific publication communicates the ideas that build the edifice of science. Scientific publication also influences the issuance of patent rights by helping to define the landscape of the prior art and obviousness criteria used in assessing the novelty of patent claims. But the different expectations from the social contract can create tensions.

Over the past 25 years, with the increasing relevance of scientific discovery for the commercial world, these tensions have increased. This is particularly true in areas such as research tools, where the discovery itself is a method or device the main commercial value of which is in furthering research. It is further complicated when an invention has the potential to be both a research tool and a therapeutic product. For example, a gene is a research tool when it is used as a probe, but it could be a therapeutic agent in a gene transfer study. Many genomic and proteomic patents are likely to have this dual character. This can be addressed by licensing the different uses under different terms (e.g., nonexclusive licensing for research and diagnostics, and exclusive by field of use for therapeutics).

Genetics is another area of conflict in which DNA is not only a tool but is something that can be abstracted simply as information and deposited in a public database, such as GenBank. In addition, DNA often cannot be invented around—that is, if one wants to study a human gene, there is effectively only one sequence to investigate. Medical research presents even further challenges with the mixing of research and medical practice in academic medical centers, especially in the conduct of clinical trials.

The EST Patent Debate

As discussed in further detail in Chapter 3, in the late 1980s and early 1990s genomic companies and universities, and NIH itself (under the leadership of then-director Bernadine Healy) began filing patents on ESTs, sparking debate in the scientific community over the public health consequences of genomic patenting and its impact on the culture of openness in science.

An EST is a small region in the active part of a gene. In the absence of genome sequence, an EST can be labeled and used as a probe to locate and isolate functional genes. Combined with a genome sequence, an EST can provide a valuable clue to the presence of a gene in the genome. Generally, EST patent applications contain broad claims, and researchers typically have identified new ESTs, guessed at the biological function of the encoded protein fragments through computerized searches of the DNA and protein databases, and then sought utility patents on the proteins on grounds of hypothetical function.

That strategy stimulated a forceful statement in 2000 by Aaron Klug and

Bruce Alberts, the presidents, respectively, of the Royal Society of London and the National Academy of Sciences in the United States. They called guessing at gene function by computerized searches of genomic data bases "a trivial matter." Its outcome might satisfy "current shareholders' interests," but it did "not serve society well." Holding that its results did not warrant patent protection, they stressed that "the human genome itself must be freely available to all humankind" (Alberts and Klug, 2000).

On the other hand, in the mid 1990s USPTO considered ESTs patentable subject matter based on a variety of utilities, such as a DNA probe. ESTs may be novel, because the sequence has not yet been published. Further, ESTs may "enable and describe at least one use, as a DNA probe."

The scientific community has expressed several concerns about the allowance of broad patent claims on ESTs, including:

- whether a DNA sequence including well-studied genes later found to contain the EST sequence would infringe the patent on that EST sequence;
- whether companies currently using gene sequences in clinical trials or those selling recombinant proteins could infringe on one or more EST patents and as a result be forced to re-engineer their gene sequence and repeat years of experiments to avoid the infringement; and
- whether industry would delay or refrain from investing in genomic research and development due to uncertainty surrounding the scope of millions of secret EST claims at USPTO that have not yet been made public either by publication of the application or issuance of the patent itself.

Many scientists believe that ESTs should not be patentable, based partly on the way they are discovered. Companies in the EST race ran genetic samples through DNA-sequencing machines that automatically identified expressed sequences but did not reveal what corresponding protein the sequence encoded or its functions. NAS President Bruce Alberts noted that, "This involves very little effort and almost no originality" (Abate, 1999). The scientific community's concerns center on the fact that ESTs have a number of immediately useful characteristics that are critical to research on hundreds of diseases. For example, an EST can be used as a label to localize that sequence on a chromosome. Because the sequence information contained in an EST is enough to distinguish one gene from all others, each EST may be used to identify the chromosomal location of its corresponding gene. The ability to identify where a particular gene is located on the chromosome is important in the detection of chromosomal mutations and corresponding disease states. Using an EST as a tool in this way could provide diagnostic tests for many diseases. Thus, restrictive patenting and licensing of ESTs potentially could throw a roadblock in the way of many pathways that investigators are taking.

In 1999, amidst growing concerns about the United States Patent and Trademark Office's (USPTO's) publicly signaled acceptance of broad claims on ESTs and other genomic inventions, then NIH Director Harold Varmus and the National Human Genome Research Institute Director Francis Collins wrote to USPTO urging the implementation of strict criteria for biotechnology patents, and specifically urging that the bar be raised for utility standards for DNA patenting so that longstanding requirements for utility would clearly apply. Since 2001, USPTO clarification of standards has become much more rigorous—that is, fewer claims are allowed.

Ultimately, through a rigorous application of existing law, USPTO did decide that it was necessary to require more than just a piece of DNA with a unique position in the genome to establish that something was useful. The applicant also must demonstrate that he or she had conceived of some practical and substantial function of that piece of DNA. However, the utility could not be something merely general and nonspecific, for example, a computer homology search indicating a similarity to a characterized protein. On the other hand, experimental data were not needed. This policy fell short for those expecting that there would be a requirement for more rigorous experimental evidence of utility. Varmus and Collins appealed to USPTO for consideration, writing:

> While we were pleased with the PTO's new stance on the utility of polynucleotides for which only generic utilities are asserted, we were very concerned with the PTO's apparent willingness to grant claims to polynucleotides for which a theoretical function of the encoded protein serves as the sole basis of the asserted utility.[9]

Currently, many patent applications directed to ESTs are pending before USPTO.

There have been persistent concerns that the patenting of an EST or indeed a gene without knowledge of its function, or with only a sketchy knowledge of its function, could preclude a product patent by some future party who discovers a much more detailed and significant functional role for that gene. Although a process patent may well be available, its value is more limited because of monitoring problems. Equally worrisome, the initial patent may block the research needed to elucidate the full role of the gene. This illustrates one feature of the patent law. A patent issued on something novel and useful can be enforced against persons seeking to use the patent for uses developed in the future. So far, USPTO has taken the statutory requirements for patents seriously and has issued few patents on ESTs. One well-known case illustrates the point, however. A firm received a patent on the gene encoding the CCR5 lymphocyte receptor without any prior knowledge of its link to HIV infection. Once the disease link was established, the

---

[9]Correspondence from Harold Varmus, NIH, and Francis Collins, NIHGR, to Q. Todd Dickinson, USPTO, dated December 21, 1999.

patentee declared its intention to enforce the patent against others, making use of the discovery in the development of any pharmaceutical to combat HIV (Johnson and Kaur, 2005).

Recent litigation has addressed the confusion related to the status of EST patents. In July 2005 a three-judge panel of the U.S. Court of Appeals for the Federal Circuit upheld USPTO in rejecting an EST patent application by Monsanto Corporation (*In re Fisher*, No. 04-1465) on the grounds that it lacked a specific, substantial, credible utility. Although the panel was split, the dissenting judge agreed that the patent should be denied—but on the grounds of obviousness rather than on utility.

Finally, now that the one-gene/one-protein postulate has proven to be overly simple (that is, one gene can encode many related variants), prior patents on genes claiming to be responsible for the production or regulation of one protein may be questioned as new discoveries are made about the many new and previously undiscovered related proteins encoded by the patented gene. Although much remains to be clarified about the appropriate standards for DNA patenting and its impact on the conduct of biological research, the burgeoning field of protein analysis raises even more consequential and distinct intellectual property policy issues. The discipline of proteomics may become an even more commercially important and active patenting arena than DNA because of its closer proximity to disease detection and therapy. Moreover, proteomics may raise novel questions of patent law that must be addressed carefully by a system that for other biological materials has evolved painfully slowly.

The Worm Model

In emerging fields, scientists often have banded together to share ideas and results before formal publication, in hopes that even more rapid scientific advances may occur. In contrast to publication, in which readers are expected to use the information without constraints, these prepublication sharing groups often impose some additional "rules of etiquette"—some restraint on the use of the information to protect the interests of the individual scientists in exchange for early access. Genetics provides several such examples, including the *Drosophila* community, in which the early adopters of this organism communicated data and materials amongst themselves before publication. The community of *C. elegans* (worm) researchers is another example and is particularly relevant, because the precedents developed can be traced directly to the data release policies of the public HGP.

C. elegans, a small free-living nematode, was selected by Sydney Brenner in the 1960s to extend his molecular genetic dissection of life from prokaryotes to animals. He considered—but discarded as too complex—the scientifically better-known and established fruit fly, *Drosophila melangaster*. Despite the obscurity of the worm, Brenner's vision of what might be learned from molecular genetic studies of this 959-celled animal soon attracted a small group of graduate students and postdoctoral fellows to Brenner's laboratory in the Medical Research Council (MRC) Laboratory of Molecular Biology, Cambridge, England.

Within the overall goal of understanding how genes specified the development and behavior of the worm, individual students applied themselves to different aspects of its development, from the early events in embryogenesis to the wiring of the nervous system. But all were united both by the overall goal and by the need to develop the tools necessary to study this recent arrival in the laboratory. Mutants of all kinds had to be isolated and the associated DNA mutations positioned on the genetic map. The genetic tools to manipulate these mutants more effectively were revised constantly. Basic methods of manipulation had to be refined. A simple but telling example is the invention of a formed, platinum wire to replace the sharpened stick used to pick individual worms under a dissecting microscope. Word quickly spread through the laboratory and soon everyone had a platinum wire "pick."

Within a few years, Brenner's postdoctoral fellows went off to establish their own laboratories, often in the United States. Feeling isolated and at a competitive disadvantage relative to their peers studying the better-established *Drosophila*, they found that communication with their colleagues in the larger Cambridge group and with other new worm laboratories proved critical for success of the next generation. This phenomenon was institutionalized by the establishment in 1975 of the *Worm Breeder's Gazette* (WBG) by Bob Edgar of Santa Cruz, an unedited compilation of contributions that was copied and sent off to all subscribers. Twice a year, laboratories submitted brief descriptions (typically one page) of new methods and exciting results, and a month later the WBG would appear. The WBG also periodically provided the community with updated genetic maps, the basic guide for all work. The information usually was shared well before any publication was planned, greatly speeding dissemination of ideas and findings. In return, readers eschewed "unfair" use of such privileged information, with fairness and individual judgment enforced by community reaction.

In addition to the WBG, the community shared stocks, mailing mutant strains around the world. Mutants were a currency of the community, providing the means of discovery. Sharing might be preceded by an explicit agreement on usage for recently obtained mutants, but mapping stocks, other tools, and published mutants were shared without constraint. These practices led eventually to the formation of a stock center, started by Bob Herman of the University of Minnesota, which centralized the activity.

It was within this community that John Sulston of the MRC Laboratory of Molecular Biology in the early 1980s began the construction of a physical map of the worm genome. The construction of the map required specialized methods carried out on thousands of clones and was considered a massive project at the time. But to be of use, the physical map had to be associated with the genetic map through genes, thereby locating the DNA fragments on the chromosomes. The community-at-large carried out this crucial activity. The central mapping laboratories (the Cambridge laboratory was joined by the Waterston laboratory at Washington University in St. Louis in the mid-1980s) provided clones to individual laboratories; in return, the laboratories identified the clones containing particular genes. As more genes were placed on the physical map, its utility became greater, rapidly escalating productivity. Again, rules of etiquette were developed to balance the interests of the community and the individual contributing laboratories. Progress on the map was communicated initially via the WBG, but as electronic media and the Internet developed, updates were provided through these media. Formal publications on the map were limited to descriptions of the methodologies and broader conclusions.

As the physical mapping project transformed into a genome sequencing project beginning in 1990, the two mapping laboratories continued their collaboration with each other and the community. The two laboratories jointly developed strategy and shared methods freely. Often the two laboratories competed in finding solutions to problems; at other times, the laboratories focused on complementary problems.

The keys to making this long-distance collaboration work were full disclosure and complete sharing of the results. In working with the community, the sequencing laboratories made the areas to be sequenced clear and submitted the results to international public databases as each segment of the genome was sequenced. But as the volume of sequence data increased along with the interest in them, it became clear that the user community could exploit the sequence data well before they were in final form. Because the public databases at the time accepted only complete sequence, the centers began posting intermediate sequence products on their Web sites. Again rules had to be developed to govern access to this prepublication data. Users were cautioned about the incomplete, imperfect nature of the sequence and were asked to consult with the centers before publication, to find out if a more complete version of the sequence was available. The centers asked to be acknowledged for providing the sequence. The users also were asked to notify the centers of any specific genes found within the sequence. This clear and practical experience in collaboration and in sharing early sequence data with the community proved invaluable when in 1996 the nascent human sequencing centers met in Bermuda to formulate plans.

The Bermuda Rules

In 1996, an international group of scientists, from both the public and private sectors, who were engaged in genomic DNA sequencing, passed a unanimous resolution, commonly referred to as the Bermuda rules, which stated that "all human genomic DNA sequence information, generated by centers funded for large-scale human sequencing, should be freely available in the public domain in

BOX C Summary of Principles of the International Strategy Meeting on Human Genome Sequencing (1996)

The following principles were endorsed by all participants. These included officers from, and scientists supported by, the Wellcome Trust, the U.K. Medical Research Council, the NIH NCHGR (National Center for Human Genome Research), the DOE (U.S. Department of Energy), the German Human Genome Programme, the European Commission, HUGO (Human Genome Organisation), and the Human Genome Project of Japan.

Primary Genomic Sequence Should Be in the Public Domain

It was agreed that all human genomic sequence information, generated by centres funded for large-scale human sequencing, should be freely available and in the public domain in order to encourage research and development and to maximise its benefit to society.

Primary Genomic Sequence Should Be Rapidly Released

- Sequence assemblies should be released as soon as possible; in some centres, assemblies of greater than 1 Kb would be released automatically on a daily basis.
- Finished annotated sequence should be submitted immediately to the public databases.

It was agreed that these principles should apply for all human genomic sequence generated by large-scale sequencing centres, funded for the public good, in order to prevent such centres establishing a privileged position in the exploitation and control of human sequence information. It was also agreed that patents should not be sought.

order to encourage research and development and to maximize its benefit to society" (see Box C). Since the sequencing phase of the publicly funded HGP began, all of the data generated by participants have been deposited in publicly available databases every 24 hours. By 2003, an essentially complete copy of the human genome sequence was posted on the Internet, with no barriers to its use, and therefore no subscription fees or other obstacles.

With the 1998 entry of Celera Genomics into the race to sequence the human genome, issues of access to the emerging data became more contentious between the public and private projects. On March 16, 2000, President Bill Clinton and U.K. Prime Minister Tony Blair issued a joint statement: "to realize the full promise of this research, raw fundamental data on the human genome, including the human DNA sequence and its variations, should be made freely available to scientists everywhere"[10]

At the beginning of the structural genomics initiative, international meetings were held to determine policies for data sharing. At a meeting of genomics and proteomics investigators held at Airlie House in Virginia in the summer of 2000, participants agreed that the coordinates for structures would be deposited in the PDB no later than six months after the completion of a structure determination.[11] The proposed delay was in direct response to the patent policies in Europe and Asia and the funding model for work being carried out on those continents. The policy for the pilot study phase of the U.S. Protein Structure Initiative (PSI) projects required data deposition within six weeks of completion of a structure determination. In Phase 2 of the PSI, the waiting period will be reduced to four weeks.

Unlike the consensus leading to the Bermuda Rules governing the HGP, general agreement on the desirability of instantaneous release of all interim data produced by structural genomics efforts was a nonstarter at Airlie House and subsequent international meetings. As for the issue of structural data release described above, much of the resistance was fueled by perceptions that patents on protein structures could generate significant financial returns within Japan and Europe. Funding agency representatives from these countries argued that without the prospect of such benefits, they would be hard pressed to make the case for funding structural genomics initiatives.

NRC Report Regarding Sharing

In 2003, NRC published *Sharing Publication-Related Data and Materials*, which stated that "Community standards for sharing publication-related data and materials should flow from the general principle that the publication of scientific information is intended to move science forward" (NRC, 2003, p. 4). The report authors argued that an author's obligation is not only to release data and materials to enable others to verify or replicate published findings but also to provide them in a form upon which other scientists can build with further research. Furthermore the report stated, "All members of the scientific community—whether working in academia, government, or a commercial enterprise—have equal responsibility for upholding community standards as participants in the publication system, and all should be equally able to derive benefits from it" (p. 4).

One of the principles embraced by the NRC committee was that if material

[10]Office of the Press Secretary. March 14, 2000. Joint statement by President Clinton and Prime Minister Tony Blair of the United Kingdom. Available at *http://clinton4.nara.gov/WH/New/html/20000315_2.html*. Accessed June 21, 2005.

[11]See *www.nigms.nih.gov/news/meetings/airlie.html*.

integral to a publication is patented, the provider of the material should make the material available under a license for research use:

> When publication-related materials are requested of an author, it is understood that the author provides them (or has placed them in an authorized repository) for the purpose of enabling further research. That is true whether the author of a paper and the requestor of the materials are from the academic, public, private not-for-profit, or commercial (for-profit) sector. Notwithstanding legal restrictions on the distribution of some materials, authors have a responsibility to make published materials available to all other investigators on similar, if not identical, terms (p. 7).

NIH Policies for Sharing and Nonexclusive Licensing

In 1999 and 2004, NIH leadership, concerned about increasingly restrictive access to research resources and data, issued guidance in two areas to encourage best practices in the scientific community. The 1999 *Principles and Guidelines for Recipients of NIH Research Grants and Contracts on Obtaining and Disseminating Biomedical Research Resources* (64 FR 72090)[12] were aspirational principles aimed at NIH-funded institutions and intended to balance the need to protect intellectual property rights with the need to disseminate new discoveries broadly. The principles apply to all NIH-funded entities and address biomedical materials, which are defined broadly to include cell lines, monoclonal antibodies, reagents, animal models, combinatorial chemistry libraries, clones and cloning tools, databases, and software (under some circumstances).[13]

Sharing Biomedical Research Resources

The 1999 principles were developed in response to complaints from researchers that restrictive terms in material transfer agreements[14] were impeding the sharing of research resources. These restrictions came both from industry sponsors and from research institutions. In the *Principles and Guidelines*, NIH urges recipient institutions to adopt policies and procedures to encourage the exchange

[12]A copy of the complete principles can be obtained at the NIH Web site at *www.nih.gov/od/ott/RTguide_final.htm.*

[13]The *Guidelines* were issued following recommendations made to the NIH Advisory Committee to the Director by a special subcommittee chaired by Rebecca Eisenberg.

[14]In the conduct of research, there is often a need to obtain compounds, reagents, test animals, cell lines, or other materials from outside individuals or entities. This is sometimes a matter of convenience—to save time and the expense of creating new research inputs—and sometimes a matter of necessity. But it also can be motivated by a desire to facilitate a research collaboration with investigators at another institution or to enable a potential corporate partner to evaluate the merit of an invention. Material transfers sometimes occur without formalities, but increasingly a material transfer agreement is used to define the rights and obligations of the parties (see Chapter 4).

of research tools, specifically: (1) minimizing administrative impediments to the exchange of biomedical research tools; (2) ensuring timely disclosure of research findings; (3) ensuring appropriate implementation of the Bayh-Dole Act; and (4) ensuring dissemination of research resources developed with NIH funds.

Four main principles are addressed in the report:

1. Ensure Academic Freedom and Publication
   a. Recipients are expected to avoid signing agreements that unduly limit the freedom of investigators to collaborate and publish.
   b. Brief delays in publication may be appropriate to permit the filing of patent applications and to ensure that confidential information obtained from a sponsor or the provider of a research tool is not inadvertently disclosed.
2. Ensure Appropriate Implementation of the Bayh-Dole Act
   a. Recipients must maximize the use of their research findings by making them available to the research community and the public, and through their timely transfer to industry for commercialization.
   b. The use of patents and exclusive licenses is not the only, nor in some cases the most appropriate, means of implementing the act. Where the subject invention is useful primarily as a research tool, inappropriate licensing practices are likely to thwart rather than promote utilization, commercialization, and public availability of the invention.
   c. Utilization, commercialization, and public availability of technologies that are useful primarily as research tools rarely require patent protection; further research, development, and private investment are not needed to realize their usefulness as research tools. In such cases, the goals of the act can be met through publication, deposit in an appropriate databank or repository, widespread nonexclusive licensing for nominal or cost-recovery fees, or any other number of dissemination techniques.
3. Minimize Administrative Impediments to Academic Research
   a. Recipients should take every reasonable step to streamline the process of transferring their own research tools freely to other academic research institutions using either no formal agreement, a cover letter, the Simple Letter Agreement of the Uniform Biological Materials Transfer Agreement (UBMTA), or the UBMTA itself.
4. Ensure Dissemination of Research Resources Developed with NIH Funds
   a. Unique research resources arising from NIH-funded research must be made available to the scientific research community.

A second section of the report, "Guidelines for Disseminating Research Resources Arising Out of NIH-Funded Research'" contains both a sample document, a UBMTA, used when transferring nonpatented materials, and sample phrases that can be used in license agreements. The goal of the NIH guidance is to simplify the process of transferring research resources from one party to another. By streamlining the process, it will help to avoid confusion about how to implement the Bayh-Dole Act properly, and will help ensure that the interests of all parties, as well as the public health, are properly balanced.

NIH Best Practices for the Licensing of Genomic Inventions

Consistent with its ongoing interest to facilitate broad access to government-sponsored research results, in 2004 NIH issued *Best Practices for the Licensing of Genomic Inventions*. This document aims to maximize the public benefit whenever Public Health Service-owned or -funded technologies are transferred to the commercial sector. In this document, NIH recommends that "whenever possible, nonexclusive licensing should be pursued as a best practice. A nonexclusive licensing approach favors and facilitates making broad enabling technologies and research uses of inventions widely available and accessible to the scientific community." The policy distinguishes between diagnostic and therapeutic applications and cautions against exclusive licensing practices in some areas.

The report considers the following to be "genomic inventions": cDNAs, ESTs, haplotypes, antisense molecules, small interfering RNAs (siRNA), full-length gene and expression products, methods, and instrumentation for the sequencing of genomes, quantification of nucleic acid molecules, detection of SNPs, and genetic modifications.

The best practice guidelines for seeking patent protection on genomic inventions depend on whether significant investment by the private sector is required to make the invention widely available. If significant investment is necessary, then patent protection should be sought. If significant investment is not necessary, however, as with many research material and research tool technologies and diagnostics, then patent protection rarely should be sought.

Regarding best practices in licensing research tools, the report recommends pursuing a nonexclusive licensing agreement whenever possible. If an exclusive license is necessary to encourage research and development by the private sector, however, then the license should be tailored to promote rapid development of as many aspects of the technology as possible. This may include limiting the field of use, specific indications, or territories the licensee has exclusive rights to develop. NIH also recommends that specific milestones be included in the licensing agreement to ensure that the technology is fully developed by the licensee. If the licensee does not meet these milestones and/or progress toward commercialization is deemed inadequate, NIH recommends that the license be modified or terminated. Additionally, whenever possible, a licensing agreement should include a provision allowing both the funding recipient and nonprofit institutions the right to use the licensed technology for research and educational purposes.

Concerns About Access to and Research Use of DNA-Based Diagnostic Tests

An area of research that has provoked the most concern, particularly among clinicians and clinical researchers, involves patents and licensing strategies for genes or partial genes associated with specific diseases. Understanding the genetics of rare and common diseases has been accelerated by the HGP, with the identification of the genes for hundreds of rare diseases, many in the past two to three years. But identifying the gene for a disease is like passing through a bottleneck. Scientists have to survey the landscape on the other side of the bottleneck to truly understand and capitalize on new opportunities for diagnosis, prevention, and treatment.

Scientists working on questions along the path between genes and function (Cho et al., 2003; Merz, 1999; Merz et al., 2002) have expressed concern about the potential restrictions that might be placed on their work if they encounter overly restrictive licensing practices by patent holders along the way. This concern is more acute in the area of diagnostic tests than it is in therapeutic product development, which clearly benefits from the protections the patent system offers during prolonged periods of research and development. The patent and licensing policies for genetic testing that followed the discoveries of the BRCA1 and 2 genes, the Canavan disease (CD) gene, and the Huntington's disease (HD) gene illustrate the many complexities of intellectual property in the area of genomics. Each case is briefly described below.

The BRCA Story

At a scientific meeting in 1990, Mary-Claire King, then a professor at the University of California, Berkeley, announced the discovery that a small region of chromosome 17 could be linked to early-onset breast cancer. This discovery was based on 15 years of research by King as well as others in this field and fueled interest in the scientific community to find the gene responsible for the high incidence of breast cancer in some families. The Breast Cancer Linkage Consortium (BCLC), an international group of scientists interested in the genetic inheritance of breast and ovarian cancer, was formed, and by pooling resources and data, scientists in the consortium were able to make discoveries more quickly than by working alone. Nevertheless, the race to the discovery of the gene was a competitive one.

In 1994, a group of scientists working under the direction of Mark Skolnick at the University of Utah announced that they had identified the gene underlying hereditary breast cancer, and named the gene BRCA1. At the time, OncorMed also performed BRCA1 diagnostic testing based on its 1997 U.S. patent on the BRCA1 consensus sequence (U.S. Patent #5,654,155). Skolnick and the University of Utah applied for and eventually were granted a U.S. patent for the gene sequence of BRCA1. They licensed the exclusive rights to Myriad Genetics, a biopharmaceutical company founded in 1991 by Mark Skolnick and others.

The studies of the BCLC also uncovered evidence that there was at least one other breast cancer gene, based on the fact that only 45 percent of familial breast cancer cases showed linkage to chromosome 17. Soon a region on chromosome 13 was identified, and the gene was localized by two groups, one led by Skolnick and Myriad Genetics and a second at the Institute for Cancer Research in the United Kingdom. BRCA1 and BRCA2 are both tumor suppressor genes whose protein products interact to control cell growth and division. Certain mutations in the BRCA1 or BRCA2 gene disrupt the regulation of growth in mammary cells, a critical step on the path to tumor formation.

Both groups filed patents in December 1995 on the second gene, termed BRCA2. CRC Technology, the technology transfer office of the Institute for Cancer Research, filed for a patent in the United Kingdom on behalf of researcher Mike Stratton, and Myriad Genetics filed for a patent in the United States. CRC Technology was granted a patent in the United Kingdom for its discovery of BRCA2 (GB 2307477), and licensed exclusively the right to a BRCA2 diagnostic test to OncorMed, a U.S. company providing genetic testing services to clients. Myriad Genetics and OncorMed were in a legal dispute over both of these patents; in 1998, the dispute finally was settled with Myriad paying OncorMed for exclusive rights to the patents. Myriad, in essence, now had a monopoly over diagnostic testing for BRCA1 and 2 familial breast cancer in the United States.

Myriad Genetics began enforcing its patent claims against certain universities, a previously rare practice. In 1999, Arupa Ganguly, of the Clinical Genetics Laboratory at the University of Pennsylvania, received a patent infringement notification from Myriad Genetics. Ganguly had developed independently a test to screen for mutations in the BRCA genes and was charging a fee to her patients to perform this test. The University of Pennsylvania was advised to cease activities related to its testing for the BRCA genes for fear of litigation by Myriad Genetics.

To curb criticism from the academic community, in 2000, Myriad Genetics negotiated an agreement with NIH so that NIH-funded researchers would receive a discount on Myriad's BRAC analysis test as long as the test was used for research purposes. The negotiated price was $1,200 per test instead of the $2,580 Myriad normally charged. In exchange for this discount, Myriad would have access to resulting research data (Hollen, 2000).

Myriad also sought patent protection on BRCA1 and BRCA2 in the European Union. In 2001, three European Union patents on BRCA1 were granted to Myriad. Myriad's first European Union patent, EP 699754, covered any methods of diagnosing a predisposition for breast and/or ovarian cancer using the normal sequence of the BRCA1 gene. Its second European Patent, EP 705903, covered 34 specific mutations of the BRCA1 gene and diagnostic methods for detecting those mutations. The third European patent, EP 705902, covered the BRCA1 gene itself, the corresponding protein, therapeutic applications of the BRCA1 gene, and diagnostic kits. In 2003, Myriad was granted a European patent on BRCA2. This patent, EP 785216, covered materials and methods used to isolate and to detect BRCA2.

In 2001 and 2002, European researchers challenged Myriad's three patents on BRCA1. When the BRCA2 patent was granted in 2003 by the European Patent Office, researchers challenged it as well. Institut Curie, a French cancer research center, and Belgian officials led the challenge along with other French and Italian research institutes. Most of the complaints fell under the categories of failure to show novelty, inventive step, industrial application, and disclosure.

In February 2004, Myriad's patent on BRCA2 was struck down because CRC Technology had filed the claim for the gene first. The European Patent Office (EPO) granted Cancer Research U.K. patent EP 868467B1 for the BRCA2 discovery. Cancer Research U.K. plans to allow free public access to its patented gene sequence.

In May 2004, Myriad's first European patent on BRCA1, EP 699754, was struck down, based on errors in the original sequence and lack of an inventive step. By the time errors were corrected by Myriad in subsequent U.S. patents, the sequence was in the public domain (Sheridan, 2004). In January 2005, two of Myriad's other BRCA1 patents met similar fates in the EPO. The scope of Myriad's claims in EP 705902 and EP 705903 was limited to probes on only the correct parts of Myriad's originally filed sequence on BRCA1 and only for testing in Ashkenazic populations. Those claims are being reviewed and may be opposed as well. These rulings by the EPO were considered a victory to those supporting free access to BRCA1 testing in Europe (Vermij, 2005).

The Canavan Disease Story

Another aspect of the controversy around genetic testing addresses the rights of patients and families whose tissue donation enable the discovery of a disease gene and eventually development of a specific test for its presence. The fight for control over the CD gene patent illustrates this debate. CD is a degenerative brain disease that irreversibly leads to loss of body control and death in children. It is a rare recessive disorder that affects about 200 children, mostly in Ashkenazi Jewish families.

Patient families were instrumental to the development of the CD genetic test. A Chicago couple with two afflicted children secured seed funding to develop a CD prenatal test, and more than 160 families provided the tissue and DNA samples that enabled discovery of the gene. Dr. Reuben Matalon at the Miami Children's Hospital (MCH) led the team that discovered the CD gene (Kaul et al., 1993). Matalon and MCH received a patent on the gene in 1997 (U.S. Patent #5,679,635).

At the time the patent was issued, the Canavan Foundation was promoting CD testing. In 1998, the American College of Obstetricians and Gynecologists recommended screening for Ashkenazi Jewish couples. Soon afterwards, MCH began enforcing its patent rights. It charged a $12.50 per test royalty and, more significantly, limited the number of laboratories that could perform the test and the number of tests performed each year. It also sought to identify one laboratory that would become the market leader and ultimately receive an exclusive license (it later abandoned this attempt). MCH came under severe criticism from patient advocacy groups and from a CD screening consortium that had been banned from performing genetic testing. In 2000, MCH initiated negotiations that involved relaxing its licensing practices and offering funds from royalties for outreach and testing in exchange for ceasing public criticism. Consortium members rejected the terms.

Also in 2000, CD families and tissue donors sued MCH; they hoped to regain control of the gene patent, testing costs, and availability. In its 2003 decision, the U.S. District Court for the Southern District of Florida dismissed several of the plaintiffs' claims—including lack of informed consent, breach of fiduciary duty, fraudulent concealment of the patent, and misappropriation of trade secrets. The judge did not dismiss the families' claim of unjust enrichment made by tissue donation, concluding that it should be litigated. MCH and the litigants decided to settle the claim; however, the settlement is sealed. Under terms of the confidential settlement, the Canavan Foundation and the families agreed to renounce any further challenges to MCH's ownership and licensing of the CD gene patent, MCH would continue to license and collect royalty fees for CD clinical testing, and license-free use of the gene in research would be allowed.

The Huntington's Disease Story

Treatment of intellectual property for the gene associated with Huntington's disease (HD) represents a counterpoint to the BRCA1/2 and CD examples. HD is an autosomal dominant disease causing involuntary movements of all parts of the body, cognitive decline, and psychiatric disturbance. It is inevitably fatal over a 10 to 20 year course. In 1979, the Hereditary Disease Foundation (HDF) organized a workshop at NIH to discuss using DNA markers to find the HD gene. HDF subsequently funded a grant to David Housman, a molecular biologist from the Massachusetts Institute of Technology, and his graduate student James Gusella to develop and use restriction fragment length polymorphisms (RFLPs) for genetic linkage analyses to locate the HD gene. Gusella continued this research at the Massachusetts General Hospital (MGH), where he began collaborating with Nancy Wexler, then at the National Institute of Neurological Disorders and Stroke. Wexler and colleagues traveled to a region in Venezuela that is home to the largest number of HD kindreds in the world, more than 18,000 people, in order to collect pedigree information and DNA for these genetic studies. The collaboration was successful; in 1983 Gusella, Housman, and colleagues discovered a RFLP marker tightly linked to the HD gene on the short arm of chromosome 4 (Gusella et al., 1983). This discovery marked the first time a disease gene of unknown chromosomal locale had been localized using anonymous DNA markers.

It would take another decade to clone the HD gene itself. To this end, scientists who discovered the HD gene marker, as well as investigators who had developed technologies for cloning genes from linked markers, in 1983 organized the Huntington's Disease Collaborative Research Group (HDCRG), under the auspices of HDF. After 10 years of collaboration by more than 50 scientists, the HD gene was isolated in 1993 (MacDonald et al., 1993). The HD gene encodes a protein called "huntingtin," the function of which remains unknown. The protein normally contains a stretch of 7 to 34 glutamines in a row. In HD, a DNA expansion causes an increase in repeats of CAG, the nucleotides encoding glutamine. In late middle age, when 40 or more CAGs occur in a row, HD almost invariably appears, usually around 30 to 40 years of age. Juvenile onset results with 60 CAG repeats or more.

When linkage was discovered in 1983, it was evident that presymptomatic and prenatal testing would be possible. These tests carry the potential to bring devastating news to HD individuals and families. Accordingly, many groups collaborated to develop detailed guidelines for HD genetic testing. In all instances, the groups included HD family members, interdisciplinary health care professionals, and research scientists. The development of counseling guidelines, including pre-, during-, and post-test counseling sessions, accompanied the development of meticulous laboratory protocols. The biotechnology company Integrated Genetics (IG), whose founders included Housman and Gusella, introduced the first HD diagnostic test in 1986. IG had been selected by MGH to carry out genetic testing, and its leadership had been included in the advisory group developing guidelines. (IG was acquired by Genzyme in 1989 and continues to do HD genetic diagnostic testing.)

Treatment of intellectual property related to the HD gene and the genetic test has been largely shaped by patient and family concerns. MGH has been granted three patents related to the HD gene. The first (U.S. Patent #4,666,828) was issued in 1987 and claimed diagnostic uses of DNA markers to detect the HD gene; Gusella was listed as the sole inventor. The others (U.S. Patents #5,686,288 and #5,693,797) were issued in 1997 and claimed huntingtin nucleic acid and protein sequences, respectively; Gusella was one of four inventors from his own laboratory. The patent application from which the 1997 patents issued was broadly written and described diagnostic and therapeutic uses of the HD gene; however, MGH did not pursue further patents claiming these methods. When Gusella and MGH filed the first application, their main concern was that the marker not be used inappropriately; they believed they might use the patent to control the testing process. Similarly, discussion of patenting the huntingtin gene usually focused on using licenses as a means to enforce testing and counseling protocols.

To date, MGH has not exerted its own patent rights or licensed the patents to others for financial gain.

The HD test is now available from more than 50 academic and commercial laboratories in the United States. It also is offered in Canada, Europe, Australia, New Zealand, Mexico, and other Latin American countries. In each region, there is no centralized licensing or pricing. There are also no complaints of the test being prohibitively expensive. (The HD genetic test is available at considerably lower cost than the BRCA tests, from $200 to $500 in the United States. The pricing of both may reflect real costs as much as business decisions. The CAG repeat test for HD involves particular primers, PCR, and accurate counting of size bands. The Myriad tests for BRCA 1 and 2 involve complete sequencing of two extremely large genes.)

The broad availability of the HD test has important ramifications for patients. First, it allows verification of test results; sending a blood sample to two different laboratories independently avoids error, which can have dire consequences. Second, the relatively low cost allows at-risk individuals to pay out-of-pocket and thus to maintain privacy with respect to insurers and employers. In the United States, many presymptomatic individuals who choose to be tested elect to pay for the test themselves. In fact, part of the genetic counseling protocol involves warning people that if they are found to have the abnormal form of the HD gene, their medical and life insurance may be jeopardized. Even individuals who are symptomatic may not want their insurance to cover the test, because a positive test implicates other family members as having a genetic risk, who then become uninsurable.

Genetic testing for HD can have profoundly catastrophic and irrevocable repercussions, as there is no treatment, prevention, or cure. Since the test became available in 1986, and where it is offered, fewer than 20 percent of eligible candidates worldwide have chosen to be tested. The psychological burden, combined with prospects for loss of privacy, insurance, and employment discrimination, have often weighed against testing. Patent-related issues, however, have not been shown to inhibit prospects for testing, in contrast to BRCA1/2.

Once the HD gene was cloned, academic and commercial laboratories interested in testing took it upon themselves to develop the proper test methodology to ensure quality control. They shared test samples representing normal and variably sized expanded alleles in order to ascertain that all the laboratories were using the same techniques and getting comparable results. This cross-checking among laboratories was also done after genetic linkage was discovered as a quality control check, but accurately counting CAG repeat numbers requires more diligence and skill. Testing quality control by sending around test samples has been done periodically ever since.

Responsibility for monitoring quality control was assumed by the academic and commercial laboratories themselves since, at least for HD and many other genetic disease tests, quality control for test performance itself currently is not monitored by any federal or state agency.

The Clinical Laboratory Improvement Amendments (CLIA) were passed by Congress in 1988 to establish "quality standards for all laboratory testing to ensure the accuracy, reliability and timeliness of patient tests." The Centers for Medicare and Medicaid Services regulates all laboratory testing (except research testing) performed on humans in the United States through CLIA regulations, but CLIA's mandate extends only to examining the physical conditions and quality control procedures of the laboratory and not to the actual performance of tests (although proficiency testing is required for many standard tests).

If the genetic test involves using a commercially available kit, FDA regulates the kit as a device and must approve the manufacturer's claims regarding the composition and performance of the kit before it can be marketed. If the test is considered a "home brew," no matter how complicated, to date neither FDA nor CLIA has opted to regulate its performance and accuracy.

In order for academic and commercial laboratories to be able to deliver an HD test result to an individual, the laboratory must be CLIA certified. As noted above, this does not mean that CLIA certifies the accuracy of specific test results, only that the laboratories in which the tests are conducted meet CLIA requirements. Many laboratories, most particularly academic laboratories that are not CLIA certified, conduct the HD test for research purposes only and these test results cannot, by law, be given to individuals, nor should they be entered into the medical record.

Gene Patents and Diagnostics

Currently, more than 1,000 genetic diseases can be diagnosed through available tests. Although some of the associated genes are free of patents, most are not. The cases of the BRCA, CD, and HD genes represent the range of situations that might apply to such genes. In a number of instances, as with HD patents, the patent holder does not enforce patent claims to prohibit testing. In some other instances, patent holders offer nonexclusive licenses for reasonable fees, making the patented tests generally available. (Following litigation, this is now the case with the CD patent.). Some gene patents, however, have been licensed exclusively, and licensees (or the original patent holders, such as Myriad) either enforce their patent rights to prevent others from performing genetic testing, or present prohibitively expensive terms under which others may perform genetic testing. Anecdotally, there are instances in which hospital laboratories performing tests on genes covered by patents have continued to offer the test until subjected to very strong pressure from the patent holder or exclusive licensee. In numerous other cases, however, hospitals have stopped offering these tests or have decided against developing genetic tests due to fear of litigation (Henry et al., 2002; Merz and Cho, 1998). (Also see further discussion in Chapter 3.)

SUMMARY

The rapid accumulation of data and information resulting from the HGP and its many spin-off projects is beginning to show movement toward clinical applications in the fields of diagnostics, therapeutics, and personalized medicine. Since the 1970s, the commercial potential of this information has been the driving force behind the growth and development of the biotechnology industry and realignment in the pharmaceutical sector. Because much of the intellectual capital in this area has resided in academic research institutions, the relationships among universities, government, and the private sector also have changed. Although these relationships have been highly beneficial, they also have generated debates about the relative roles of government and industry in supporting and promoting science, particularly with regard to open access to information and the sharing of research resources. Also at issue is what can or should be patented and whether there is an obligation to the public health to ensure that clinical and research access to valuable discoveries is not unduly restricted.

3

The U.S. Patent System, Biotechnology, and the Courts

The basis of the American patent system is found in Article I, Section 8 of the U.S. Constitution, which empowers Congress "to promote the Progress of Science and useful arts, by securing for limited Times to . . . Inventors the exclusive Right to their . . . Discoveries." Congress enacted the first patent statute in 1790 and amended it in 1793. The 1793 amendment defined, in language written by Thomas Jefferson, what was patentable: "any new and useful art, machine, manufacture, or composition of matter, or any new or useful improvement thereof." Jefferson's phrasing remains at the core of the U.S. patent code, except for the replacement of the 18th-century word "art" with the modern equivalent "process" in a 1952 congressional overhaul of patent law.

This chapter addresses intellectual property issues in the context of genomics and proteomics, focusing on patent law and interpretation—specifically, fields of activity, applicable law, and limitations on licensing and enforcement.

FIELDS OF ACTIVITY

The practice of patenting genes is at least as old as the biotechnology industry, providing a quarter century's worth of legal precedents on the application of patent law to genetics and genomics. Inevitably, however, legal developments trail behind scientific developments, particularly in rapidly advancing fields. As the underlying science has advanced, research strategies and business models have become more diverse, generating patents that play different roles in the economy of biomedical research and practice than those of the early days of the biotechnology industry. Although some legal issues are now reasonably well settled, new and unresolved issues have come into view.

Beginning in the 1970s, firms sought and obtained patents on newly cloned genes encoding therapeutic proteins (Eisenberg, 1990). These early patents typically claimed an "isolated and purified" DNA sequence corresponding to the amino acid sequence for the protein, along with recombinant materials incorporating that DNA sequence for use in making the protein in cultured cells.[1] As a legal matter, the courts and the United States Patent and Trademark Office (USPTO) treated these inventions as chemicals or "composition of matter,"[2] a characterization that provided an extensive body of precedent that could be consulted to establish the ground rules for patents in this emerging field. Having long ago decided that chemicals isolated from nature through human intervention were eligible for patent protection,[3] the courts and USPTO had little difficulty allowing patents on newly isolated genes.

The gene-patenting pioneers in the new biotechnology firms of the 1980s saw themselves as high-technology drug developers, and in their search for a viable business model for therapeutic protein development, they emulated the patent strategies of major pharmaceutical firms. Patents on the genes that encoded therapeutic proteins secured exclusive franchises to manufacture these products. Such patents have been the focus of numerous judicial opinions concerning the requirements for patent protection,[4] priority of invention,[5] and determinations of infringement.[6] The judicial opinions that resolve these disputes provide most of the existing legal precedent involving the patenting of DNA.

Following the first wave of patents on genes encoding therapeutic proteins, the development of new tools and techniques for detecting genetic differences among individuals enabled researchers to bypass the stages of protein isolation and characterization and to identify directly the genes associated with diseases

---

[1]See, e.g., U.S. Patent No. 4,757,006 (July 12, 1988), which claims, *inter alia*: 1. An isolated recombinant vector containing DNA coding for human factor VIII:C, comprising a polydeoxyribunucleotide having the [following] sequence: 4. A nonhuman recombinant expression vector for human factor VIII:C comprising a DNA segment having the [following] sequence: 5. A transformed non-human mammalian cell line containing the expression vector of claim 4.

[2]See, e.g., *Amgen v. Chugai Pharmaceutical Co.*, 927 F.2d 1200, 1206 (Fed. Cir.), *cert. denied sub nom. Genetics Institute v. Amgen*, 502 U.S. 856 (1991) ("A gene is a chemical compound, albeit a complex one . . .").

[3]E.g., *Parke-Davis & Co. v. H.K. Mulford & Co.*, 189 F. 95 (S.D.N.Y. 1911) (adrenaline); *Kuehmsted v. Farbenfabriken*, 179 F. 701 (7th Cir. 1910), *cert. denied*, 220 U.S. 622 (1911) (prostaglandins); *Merck & Co. v. Olin Mathieson Corp.*, 253 F.2d 156 (4th Cir. 1958) (vitamin B12).

[4]See, e.g., *In re Deuel*, 51 F.3d 1552 (Fed. Cir. 1995) (nonobviousness); *Regents of the University of California v. Eli Lilly*, 119 F.3d 1559 (Fed. Cir. 1997) (nonobviousness); *Genentech v. Novo Nordisk*, 108 F.3d1361 (Fed. Cir. 1997) (enablement); *Eli Lilly v. Genentech*, 119 F.3d 1567 (Fed. Cir. 1997) (written description).

[5]See, e.g., *Fiers v. Revel*, 984 F.2d 1164 (Fed. Cir. 1993).

[6]See, e.g., *Scripps Clinic & Research Found. v. Genentech*, 927 F.2d 1565 (Fed. Cir. 1991); *Genentech v. Wellcome Foundation*, 29 F.3d.

(or disease susceptibilities) through positional cloning (Collins, 1995). These genetic discoveries had immediate value as diagnostic products. They also were useful as research tools in the development of therapeutic products, but the relationship between gene and therapeutic product typically was less straightforward than it had been for the first generation of biotechnology products. Patents on these discoveries, although similar in form to patents on genes encoding therapeutic proteins, played a different and less familiar role in the biomedical community, setting the stage for conflict between people and institutions that had barely taken note of the first generation of gene patents. Professional societies of doctors and clinical geneticists in particular have been outspoken critics of disease gene patents, especially when they are the subject of exclusive licenses to perform DNA diagnostic tests.[7] They argue that patent-based restrictions regarding who may perform genetic tests interfere with the practice of medicine and prevent other laboratories from identifying and validating new mutations. These arguments are particularly compelling to doctors and geneticists working in academic medical centers that are equipped to administer "home brew" genetic diagnostic tests themselves in pursuit of a mixed mission of treatment and research.

The advent of high-throughput DNA sequencing marked another important turning point in the history of genomic patents. By generating large amounts of DNA sequence information in advance of understanding the functions or disease relevance of particular sequences, high-throughput sequencing raised the possibility of obtaining patents on "upstream" genetic discoveries that—although potentially possessing patentable utility—were still far removed from developed products. As discussed in Chapter 2, the announcement of the filing by NIH of patent applications on the first expressed sequence tags (ESTs) identified by Craig Venter in a National Institutes of Health (NIH) laboratory set off controversy in the scientific community (Dickson, 1993; Milstein, 1993), although research scientists previously had expressed little concern about the patenting of genes encoding therapeutic proteins.

If the DNA sequence discoveries that were claimed in the provocative NIH patent filings encoded therapeutic proteins or were relevant to particular diseases, no one could have known it at the time. The most obvious value of ESTs was not

[7]See, e.g., Association for Molecular Pathology, AMP Position on Patenting of Genetic Tests (Dec. 17, 1999), posted on the Internet at *www.ampweb.org/patent.htm*; American College of Medical Genetics, Position Statement on Gene Patents and Accessibility of Gene Testing (Aug. 2, 1999), posted on the Internet at *www.faseb.org/genetics/acmg/pol-34.htm*; *American Medical Association, H-140.944 Patenting the Human Genome, posted on the Internet at www.ama-assn.org/apps/pf_online?f-n=browse&doc=policyfiles/HOD/H-14C*; Academy of Clinical Laboratory Physicians and Scientists, Resolution: Exclusive Licenses for Diagnostic Tests (approved by the ACLPS Executive Council June 3, 1999), posted on the Internet at *depts.washington.edu/lmaclps/license.htm*; College of American Pathologists, Gene Patents Detrimental to Care, Training, Research, posted on the Internet at *www.cap.org/html/advocacy/issues/Issue_Genepat.html.*

the speculative value that particular gene fragments might have for therapeutic or diagnostic uses, but the immediate value that collections of such sequences offered for use in gene discovery. With this shift, patenting genes began to look less like patenting end products and more like patenting research tools. Scientists argued that the progress of biomedical research would be better served by making the human genome freely available than by permitting its balkanization through patent claims and restrictive licensing agreements.[8] This opposition became more vehement as EST sequencing moved from NIH to the private sector and as terms of access to privately held EST databases were set beyond the reach of many academic institutions (Eisenberg, 1996).

The EST patenting controversy arose during a period of rapid transition in the culture of academic biomedical research from a tradition of open science to a more restrictive, proprietary enterprise. Although by this point many academic scientists had begun patenting their own inventions and licensing them through their universities to private firms, they still enjoyed relatively free access to scientific information and methods for use in their own research. The first patent filings on the results of high-throughput DNA sequencing coincided with a broader trend in the biomedical research community to claim intellectual property rights in research tools and to assert these rights against academic researchers (NIH Working Group on Research Tools, 1998).

In addition to raising concerns about the patenting of research tools, the EST controversy highlighted the character of genomic discoveries as *information*, as distinguished from tangible molecules. Much of the value of ESTs lay in databases, rather than in tangible materials stored in a wet laboratory. From this perspective, ESTs were an early harbinger of an aspect of genomics and proteomics that has continued to be problematic for the patent system. The sequencing of genomes, the identification of polymorphisms and haplotypes, the development of gene expression profiles, and the determination of protein structures all involve the creation of valuable information resources and the analysis of information on a large scale.

It is not always obvious how to use the patent system to capture the value of these information resources. The courts and USPTO initially resisted the extension of patent protection to information technology,[9] although a series of deci-

---

[8]Opposition to gene patenting from the scientific community has become more qualified over time, as scientific institutions have sought to establish ground rules that would limit patent protection to well-characterized genes while withholding patents on gene fragments and sequences whose function has not been established. These more qualified views have recently been set forth in comments on proposed USPTO guidelines on the utility and written description requirements for patent protection, posted on the Internet at *www.usUSPTO.gov/web/offices/com/sol/comments/utilguide/index.html* and *www.usUSPTO.gov/web/offices/com/sol/comments/utilitywd/index.html.*

[9]*Gottschalk v. Benson*, 409 U.S. 63 (1972); Parker v. Flook, 437 U.S. 584 (1978).

sions of the Federal Circuit over the past decade have attenuated this once robust exclusion.[10]

From a commercial perspective, genomic and proteomic information resources are an important part of platforms for the development of future diagnostic and therapeutic products, and the developers of these resources have pursued a variety of patent claiming and licensing strategies that would permit them to share in the potential bounty of these future products. One controversial strategy is "reach-through licensing" of platform technologies in exchange for either royalties on future products that would not otherwise infringe the upstream patent (reach-through royalties) or a promise of an exclusive or nonexclusive license under future patents on inventions to be made by the user (grant-backs) (Eisenberg, 2001).

The platform technology may be a patented research tool, or it may be an unpatented database that is accessible only under the terms of a database access agreement. Owners of platform technologies often seek reach-through royalties when the user is a pharmaceutical firm or other institution engaged in product development.[11] Grant-backs are more typical when the user itself is a nonprofit institution, such as a university, that does not develop products but nonetheless generates further intellectual property.[12] Both are primarily contract strategies, in which the role of patent law is largely limited to determining background rights. A more powerful strategy for allowing developers of research tools to capture the value that their tools contribute to future product development is called "reach-through claiming."[13] This strategy focuses on the patent itself rather than on license terms, using claim language that is broad enough to cover future products directly. Examples of reach-through claims are those directed to agonists or antagonists of a disclosed receptor, to methods of treatment involving administering to a patient a compound identified through a disclosed screening method, or to products designed to fit within a binding site on a protein for which crystal coordinates have been disclosed. Some of these claiming strategies raise unresolved legal issues, although it appears that the "written description" requirement would pose a significant obstacle to the patentability of such claims.

---

[10]*State St. Bank & Trust v. Signature Financial Group*, 149 F.3d 1368 (Fed. Cir. 1998), *cert. denied*, 119 S. Ct. 851 (1999); *AT&T Corp. v. Excel Communications, Inc.*, 172 F.3d 1352 (1999).

[11]*Id.* at 243. See, e.g., *Bayer AG v. Housey Pharmaceuticals, Inc.*, 228 F. Supp. 2d 467 (2002).

[12]*Bargaining Over Research Tools, supra* note 17, at 244.

[13]For a description of reach-through claiming and a careful analysis of the doctrinal issues it raises under the patent laws in the U.S., the European Union, and Japan, see *Report on comparative study on biotechnology patent practices theme: Comparative study on "reach-through claims"* (Trilateral Project B3b, posted on the Internet at *www.USPTO.gov/web/tws/b3b_reachtrhougyh.doc*, hereinafter "Comparative Study").

APPLICABLE LAW

In order to get a patent, an applicant must claim an invention that falls within a patent-eligible subject matter. The invention must be new, useful, and nonobvious in light of the prior art. The patent application must satisfy certain disclosure requirements, including a written description of the invention, an enabling disclosure that would allow a person of ordinary skill in the field to make and use the invention without undue experimentation, and disclosure of the best mode contemplated by the inventor of carrying out the invention. A patent gives its owner the right to exclude anyone else from making, using, selling, offering to sell, or importing the invention during the patent term, subject to limited exceptions. Patent owners generally enjoy considerable discretion to deploy these exclusionary rights as they see fit, although some licensing practices may violate antitrust laws or constitute "patent misuse." For inventions made in the course of government-sponsored research, the government retains the right to practice the invention or to authorize others to practice it for governmental purposes, as well as "march-in" rights to grant licenses under the patent if necessary to get the invention developed.

Patent Eligibility

Do discoveries in genomics and proteomics fall within the range of subject matter that the patent system protects? On its face, the Patent Act extends protection to "any new and useful process, machine, manufacture or composition of matter," without explicit subject matter exclusions.[14] But in the past the courts and USPTO have sometimes seemed to endorse exclusions from patent eligibility for certain categories of inventions, including medical and surgical techniques,[15] plants,[16] agricultural methods,[17] mathematical algorithms,[18] and products and

[14]35 U.S.C. § 101.

[15]*Morton v. New York Eye Infirmary*, 17 F. Cas. 879 (No. 9865) (S.D.N.Y. 1862) (holding ineligible for patent protection method of performing surgery by applying ether to render patient insensitive to pain); Ex parte Brinkerhoff, 24 Comm'n MS Decision 349 (1883) (holding that "the methods or modes of treatment of physicians of certain diseases are not patentable"). But cf. *Smith & Nephew v. Ethicon*, 54 USPQ2d 1888, 1889 (D. Ore. 1999) (claiming "a method of attaching tissue to bone by using a resilient suture anchor which is pressed into a hole in the bone"); *Catapano v. Wyeth Ayerst Pharmaceuticals*, 88 F. Supp.2d 27, 28 (E.D.N.Y. 2000) (claiming a method of treating a human patient to effect the remission of AIDS).

[16]*Ex parte Latimer*, 1889 Comm'n Dec. 13 (1889) (holding ineligible for patent protection a claim to "cellular tissues of the Pinus australis" tree separated from "the silicious, resinous, and pulpy parts of the pine needles and subdivided into long, pliant filaments adapted to be spun and woven"). But cf. *J.E.M. Ag Supply v. Pioneer Hi-Bred International*, 534 U.S. 124 (2001) (holding plants eligible for patent protection).

[17]*Wall v. Leck*, 66 F. 552 (9th Cir. 1895) (invalidating patent on a process of fumigating citrus trees in the absence of light).

[18]*Gottschalk v. Benson*, 409 U.S. 63 (1972).

phenomena of nature.[19] These exclusions have been viewed skeptically by the Court of Appeals for the Federal Circuit (Federal Circuit) and by its predecessor, the Court of Customs and Patent Appeals,[20] and by now most have been repudiated.[21] In the landmark case of *Diamond* v. *Chakrabarty,* the U.S. Supreme Court held that the broad language of the patent statute indicates an expansive scope of eligible subject matter that includes "anything under the sun that is made by man."[22] USPTO has issued and the courts have upheld patents on tangible DNA and proteins in forms that do not occur in nature as new "compositions of matter." Patents have thus issued on "isolated and purified" DNA sequences and proteins and on DNA sequences that have been spliced into recombinant vectors or introduced into recombinant cells of a sort that do not exist in nature.[23] This is consistent with longstanding practice, even prior to the advent of modern biotechnology, of allowing patents to be issued on isolated and purified chemical products that exist in nature only in an impure state, when human intervention has made them available in a new and useful form.[24]

More recent advances in genomics and proteomics raise somewhat different issues concerning the patent eligibility of information, as distinguished from tangible new products and processes. Older cases have excluded from patent protection "scientific truths" and "abstract ideas."[25] The sequencing of genomes, the identification of polymorphisms and haplotypes, the development of gene expression profiles, and the determination of protein structures all provide valuable scientific information that arguably falls within these exclusions, to the extent

---

[19]*Funk Bros. Seed v. Kalo. Inoculant Co.*, 333 U.S.127 (1948).

[20]Congress created the Court of Appeals for the Federal Circuit in 1982, consolidating intermediate appellate jurisdiction over patent law matters in a single court that would hear appeals from decisions of USPTO and decisions of the Federal District Courts in patent cases. Federal Courts Improvement Act of 1982, Pub. L. No. 97-164, 96 Stat. 25. An important goal was to bring about greater uniformity and consistency in interpretations of the patent laws.

[21]E.g., *State St. Bank & Trust v. Signature Financial Group*, 149 F.3d 1368 (Fed. Cir. 1998), *cert. denied*, 515 U.S. 1093 (1999); *AT&T Corp. v. Excel Communications, Inc.*, 172 F.3d 1352 (1999); *Pioneer Hi-Bred Int'l v. J.E.M. Ag Supply*, 200 F.3d 1374 (Fed. Cir. 2000), aff'd sub nom. *J.E.M. Ag Supply v. Pioneer Hi-Bred Int'l*, 534 U.S. 124, 130 (2001).

[22]*Diamond v. Chakrabarty*, 447 U.S. 303 (1980).

[23]*Amgen, Inc. v. Chugai Pharmaceutical Co.*, 13 U.S.P.Q.2d (BNA) 1737 (D. Mass. 1990). ("The invention claimed in the '008 patent is not as plaintiff argues the DNA sequence encoding human EPO since that is a nonpatentable natural phenomenon 'free to all men and reserved exclusively to none.' . . . Rather, the invention as claimed in claim 2 of the patent is the "purified and isolated" DNA sequence encoding erythropoietin.")

[24]See, e.g., *Merck & Co. v. Olin Mathieson Chemical Corp.*, 253 F.2d 156 (4th Cir. 1958) (upholding the patentability of purified Vitamin B-12). See also *In re Bergy,* 596 F.2d 952 (CCPA 1979) (upholding patentability of isolated and purified microorganism), vacated and remanded with directions to dismiss as moot sub nom. *Diamond v. Chakrabarty* 444, U.S. 1028 (1980).

[25]See, e.g., *LeRoy v. Tatham*, 55 U.S. 156, 175 (1853)("A principle, in the abstract, is a fundamental truth; an original cause; a motive; these cannot be patented, as no one can claim in either of them an exclusive right."); *Mackay Co. v. Radio Corp.*, 306 U.S. 86, 94 (1939) ("While a scientific truth, or

that the exclusions remain good law, if the patent claims read beyond the materials themselves and attempt to define the invention in such a way that the use of information would be an act of patent infringement. Recent decisions concerning the patentability of computer-implemented inventions may provide more guidance than prior decisions about the patentability of discoveries in the life sciences and in predicting the patentability of informational inventions in genomics and proteomics.

The overall trend of decisions in the Federal Circuit is toward the expansive interpretation of the scope of patent-eligible subject matter—even for categories of inventions that prior decisions seemed to exclude from the protection of the patent statute—in order to make the patent system "responsive to the needs of the modern world."[26] The most conspicuous recent example of this trend was the 1998 decision in *State Street Bank & Trust* v. *Signature Financial Group*[27] upholding the patentability of a computer-implemented accounting system for managing the flow of funds in partnerships of mutual funds that pool their assets. Despite the fact that USPTO had been issuing similar patents for many years, it was argued that this invention fell within some previous judicial limitations that arguably excluded mathematical algorithms[28] and business methods[29] from patent protection. The Federal Circuit minimized the first of these limitations,[30] holding that it excluded from patent protection only "abstract ideas constituting disembodied concepts or truths that are not 'useful,'"[31] and repudiated the second, insisting that "[t]he business method exception has never been invoked by this court, or [its predecessor], to deem an invention unpatentable," and that other courts that had appeared to apply the business method exception always had other

---

the mathematical expression of it, is not a patentable invention, a novel and useful structure created with the aid of knowledge of scientific truth may be."); *Gottschalk v. Benson*, 409 U.S. 63 (1972) ("It is conceded that one may not patent an idea. But in practical effect that would be the result if the formula for converting BCD numerals to pure binary numerals were patented in this case."); *Diamond v. Chakrabarty* ("Einstein could not patent his celebrated law that $E=mc^2$; nor could Newton have patented the law of gravity. Such discoveries are 'manifestations of . . . nature, free to all men and reserved exclusively to none.'"); *Dickey-John Corp. v. International Tapetronics Corp.*, 710 F.2d 329 (7th Cir. 1983) ("Yet patent law has never been the domain of the abstract—one cannot patent the very discoveries that make the greatest contributions to human knowledge, such as Einstein's discovery of the photoelectric effect, nor has it ever been considered that the lure of commercial reward provided by a patent was needed to encourage such contributions. Patent law's domain has always been the application of the great discoveries of the human intellect to the mundane problems of everyday existence.")

[26]*AT&T Corp. v. Excel Communications, Inc.*, 172 F.3d 1352 (1999).

[27]149 F.3d 1368 (Fed. Cir. 1998), *cert. denied,* 119 S. Ct. 851 (1999).

[28]See, e.g., *Gottschalk v. Benson*, 409 U.S. 63 (1972); *Parker v. Flook*, 437 U.S. 584 (1978).

[29]*Hotel Security Checking Co. v. Lorraine Co.*, 160 F. 467 (2d Cir. 1908).

[30]The exclusion of mathematical algorithms from patent protection had already been substantially restricted by prior decisions of the Federal Circuit. See, e.g., *In re Alappat*, 33 F.3d 1526 (1996).

[31]149 F.3d at 1373.

grounds for arriving at the same decision.[32] Rather than seeing the language of §101 of the Patent Act as a significant limitation on the types of advances that might qualify for patent protection, the Federal Circuit characterized the statutory language as a "seemingly limitless expanse," subject only to three "specifically identified . . . categories of unpatentable subject matter: 'laws of nature, natural phenomena, and abstract ideas.'"[33]

So far, USPTO has declined to issue patents on data as such, as have its counterparts in Europe and Japan. USPTO's Examination Guidelines for Computer-Implemented Inventions[34] distinguish between "functional descriptive material" (such as "data structures and computer programs which impart functionality when encoded on a computer-readable medium") and "non-functional descriptive material" (such as "music, literary works and a compilation or mere arrangement of data [which] is not structurally and functionally interrelated to the medium but is merely carried by the medium").[35]

A 2002 report on a trilateral comparative study by the European Patent Office, the Japan Patent Office, and USPTO ("2002 trilateral report") considers the patentability of protein three-dimensional structure-related claims under the laws administered by each of those offices.[36] Each of the three offices concluded that hypothetical claims to computer models of proteins generated with atomic coordinates, data arrays comprising the atomic coordinates of proteins, computer-readable storage medium encoded with the atomic coordinates, and databases

---

[32]*Id.* At 1375-76.

[33]*AT&T Corp. v. Excel Communications, Inc.*, 172 F.3d 1352 (Fed. Cir. 1999), citing *Diamond v. Diehr*, 450 U.S. 175 (1981).

[34]61 Fed. Reg. 7478 (Feb. 28, 1996), posted on the Internet at *www.usUSPTO.gov/web/offices/com/hearings/software/analysis/computer.html.*

[35]The focus on functional relationship between data and substrate echoes language from *In re Lowry*, 32 F.3d 1579 (Fed. Cir. 1994), in which the Federal Circuit upheld the patentability of a data structure for storing, using, and managing data in a computer memory. In that case, the Board of Patent Appeals had reversed the examiner's rejection of the claims under 35 U.S.C. § 101 as claiming non-statutory subject matter, and the issue of patentable subject matter was therefore not properly before the court on appeal. Nonetheless, in its analysis of the remaining issues of patentability under 35 U.S.C. §§ 102 and 103, the court drew a distinction between claiming information content and claiming a functional structure for managing information: "Contrary to the USPTO's assertion, Lowry does not claim merely the information content of a memory. Lowry's data structures, while including data resident in a database, depend only functionally on information content. While the information content affects the exact sequence of bits stored in accordance with Lowry's data structures, the claims require specific electronic structural elements which impart a physical organization on the information stored in memory. Lowry's invention manages information. As Lowry notes, the data structures provide increased computing efficiency."

Id. at 1583. See also *In re Warmerdam*, 33 F.3d 1354, 1361 (Fed. Cir. 1994) (descriptive material per se is not patent eligible subject matter).

[36]Trilateral Project WM4, Comparative studies in new technologies, Report on comparative study on protein 3-dimensional (3-D) structure related claims (Nov. 4-8, 2002).

encoded with candidate compounds that had been electronically screened against the atomic coordinates of proteins were not patent-eligible subject matter. The analysis of USPTO emphasized that the subject matter of each of these hypothetical claims was "nonfunctional descriptive material" and therefore "an abstract idea."

Although the views of USPTO and its foreign counterparts are of enormous practical importance in determining what gets patented, neither the USPTO guidelines nor the 2002 trilateral report has the status of binding legal authority. If a disappointed patent applicant appeals the rejection of claims covering atomic coordinates for proteins to the Federal Circuit, that court could reverse the rejection.

To the extent that the exclusion of "nonfunctional descriptive material" from patent protection rests on the intangible nature of information, it may prove unstable in the face of recent Federal Circuit decisions that have de-emphasized the need for structural physical limitations in claims for computer-implemented inventions. This shift in emphasis is particularly apparent in *AT&T* v. *Excel Communications*,[37] in which the court explicitly declined to focus on the "physical limitations inquiry" that had played a central role in distinguishing between unpatentable mathematical algorithms and patentable computer-implemented inventions in its prior decisions. Instead, the court asked "whether the mathematical algorithm is applied in a practical manner to produce a useful result."[38] This approach seems to merge the issue of patent eligibility with the issue of utility. It is unclear, however, whether the court would take the further step of opening the door to patent claims to information itself so long as it is "useful," contrary to the time-honored understandings of the subject matter categories that are eligible for patent protection.[39]

---

[37]172 F.3d 1352 (Fed. Cir. 1999).

[38]*Id.* at 1358.

[39]After the committee completed its deliberations, the issue of patent eligibility was raised in two settings. First, on October 26, 2005, the PTO issued new Interim Guidelines for Examination of Patent Applications for Patent Subject Matter Eligibility, see *http://www.uspto.gov/web/offices/pac/dapp/ogsheet.html*. If adopted, these guidelines would most directly affect the patenting of computer-related and business method patents. More pertinent to the issues raised here, the Supreme Court granted certiorari in *Laboratory Corporation of America Holdings v Metabolite Laboratories Inc.*, 370 F.3d 1354 (Fed. Cir. 2004), (Supreme Court 2005). WL 2838583 (U.S. Oct. 31, 2005) (No. 04-607). The case raised several challenges to the validity of a patent claiming methods for detecting cobalamin (vitamin B_{12}) or folate deficiency by testing for elevated levels of total homocysteine. However, the Court limited the scope of its review to the following question:

> Whether a method patent setting forth an indefinite, undescribed, and non-enabling step directing a party simply to "correlat[e]" test results can validly claim a monopoly over a basic scientific relationship used in medical treatment such that any doctor necessarily infringes the patent merely by thinking about the relationship after looking at a test result.

Because the grant of certiorari is limited, it is difficult to know how far the decision of the Court might reach. Nonetheless, the case provides a vehicle for determining when diagnostic tests will be considered patent-eligible.

In October 2005, against the solicitor general's advice, the Supreme Court announced that it will hear arguments in a case that could narrow the scope of patentability in the United States. In granting certiorari in *Laboratory Corporation of America Holdings v Metabolite Laboratories Inc.*, the Court chose to address just one of the three questions presented by the petitioner and will focus only on the area of medical diagnostic and treatment patents. The question the Court will answer is "Whether a method patent setting forth an indefinite, undescribed, and non-enabling step directing a party simply to 'correlat[e]' test results can validly claim a monopoly over a basic scientific relationship used in medical treatment such that any doctor necessarily infringes the patent merely by thinking about the relationship after looking at a test result."

The case involves a diagnostic test carried out by doctors. Metabolite's patent covers tests to determine homocysteine levels in the body, and according to Metabolite, additionally covers scientific relationship between homocysteine and vitamin B deficiencies. Thus, argues Metabolite, the patent claims all forms of correlating test results, such as a doctor seeing low homocysteine results and determining low vitamin B levels.

The solicitor general's brief had argued for denial of certiorari on the grounds that the history of the case and the facts of the dispute were not suitable to address the broader question of patent eligibility. The Court's acceptance of the case, however, has set the stage for what could be a significant revision of patentability.

Utility

Another limitation on what may be patented that has been important in genomics is the utility requirement. The utility requirement straddles two statutory provisions in U.S. law: § 101, which defines patent eligible subject matter as "any new and *useful* process, machine, manufacture and composition of matter," and § 112, which requires that a patent applicant disclose "the manner and process of making and *using*" the invention.[40] In most fields of technology, the utility requirement does little work, because few people would want to patent a useless invention and few would care if they did, but the requirement plays a more important role in chemistry and biotechnology. This is because the course of discovery in these fields typically involves the identification of products first, followed by screening or testing for uses later. The need to describe and enable utility thus potentially defers the time when inventions in these fields may be

[40]The requirement also finds authority in the language of Article 1, section 8, clause 8 of the U.S. Constitution, which authorizes Congress to enact patent legislation for a specified purpose: "To promote the Progress of Science and *useful* Arts . . . ." [emphasis added].

patented. Patent law promotes the early filing of patent applications through novelty and statutory bar standards that put dilatory applicants at risk of losing patent protection entirely.[41] This motivates inventors to file patent applications on new molecules as soon as possible, raising the issue of how much of a description of utility is necessary to get a patent on a product whose practical uses may not yet be well understood or definitively established.

The Supreme Court articulated a relatively strict utility standard in its 1966 decision in *Brenner* v. *Manson*, requiring that a patent applicant show that the invention has "specific benefit in currently available form." The court justified this strict approach by noting that "a patent is not a hunting license. It is not a reward for the search, but compensation for its successful conclusion." But the standard has proven difficult to apply in a predictable fashion as technology advances. More recent decisions of the Federal Circuit have appeared to be less strict about the utility requirement than Supreme Court precedent dictates.[42]

In the early 1990s, there was a widespread perception in the biotechnology patent community that USPTO examiners were applying an unduly strict utility standard, requiring patent applicants to submit the sort of data that would satisfy the Food and Drug Administration (FDA) that a product was safe and effective before they would allow patents on therapeutic inventions. Patent applicants appealed, and the Federal Circuit eventually reversed USPTO in the case of *In re Brana*,[43] holding that "[u]sefulness in patent law, and in particular in the context of pharmaceutical inventions, necessarily includes the expectation of further research and development." The court admonished USPTO that an applicant's assertion of utility is presumptively correct unless based on implausible scientific principles, and the burden is initially on USPTO to provide evidence showing that someone of ordinary skill in the art would reasonably doubt the asserted utility before it enters a rejection for lack of utility. This suggests that the Federal Circuit is inclined to accept plausible speculation about how to use the invention in satisfaction of the utility requirement, and that inventors should not ordinarily have to await extensive data collection before filing for patents. But, if an inventor's plausible speculation proves to have been incorrect, such a patent may

---

[41]A patent application is barred under § 102(b) of the Patent Act if the inventor fails to file within one year of first publication or other public use of the invention. Moreover, the dilatory applicant who keeps the invention secret risks losing priority to another applicant who subsequently claims the same molecule if he is deemed to have "abandoned, suppressed, or concealed" the invention. 35 U.S.C. § 102(g).

[42]See, e.g., *Juicy Whip v. Orange Bang*, 185 F.3d 1364 (Fed. Cir. 1999) (fact that invention may deceive some members of the public does not deprive it of utility); *In re Brana*, 51 F.3d 1560 (Fed. Cir. 1995) (reversing rejection of patent application on compounds for use as antitumor substances, notwithstanding absence of data from human clinical trials indicating efficacy).

[43]51 F.3d 1560 (Fed. Cir. 1995).

ultimately be invalid for failure to disclose and enable an operable entity. Around the same time as the Federal Circuit's decision in *In re Brana*, USPTO held public hearings on the utility requirement and issued revised Utility Examination Guidelines echoing the Federal Circuit's call for examiners to be more cautious about entering utility rejections consistent with the statutory standard.[44]

These 1995 guidelines were soon superseded by 2001 guidelines after it became clear that examiners had become too lax, effectively raising the utility standard once again for biotechnology inventions.[45] Patent applications on ESTs presented the standard of patentable utility in a somewhat different light, raising issues that had not been resolved by prior cases. The scientific community urged USPTO to use the utility requirement to limit the patenting of gene fragments of unknown function.[46] After providing extensive opportunity for public input, and acknowledging that the utility standard should apply equally to all inventions regardless of technology, USPTO fortified its articulation of the utility standard for biotechnology examiners in new Utility Examination Guidelines. These guidelines directed examiners to apply Supreme Court precedent and to require that the application assert "a specific and substantial utility that is credible."[47] Using a genomics example, USPTO elaborated that "a claim to a polynucleotide whose use is disclosed simply as a 'gene probe' or 'chromosome marker' would not be considered to be *specific* in the absence of a disclosure of a specific DNA target."[48] On the other hand, in its responses to comments, USPTO noted that "the utility of a claimed DNA does not necessarily depend on the function of the encoded gene product," but might, for example, be established by a credible assertion that "it hybridizes near a disease-associated gene or it has a gene regulating activity."[49] USPTO explicitly declined to adopt a *per se* rule against assertions of utility based upon homology to prior art sequences, citing *In re Brana* and noting the absence of "scientific evidence that homology-based assertions of utility are inherently unbelievable or involve implausible scientific principles."[50]

---

[44]U.S. USPTO, *Utility Examination Guidelines*, 60 Fed. Reg. 36,263 (July 14, 1995).

[45]U.S. USPTO, *Utility Examination Guidelines*, 66 Fed. Reg. 1092 (Jan. 5, 2001).

[46]See Public Comments on the United States Patent and Trademark Office "Revised Interim Utility Examination Guidelines," 64 Fed. Reg. 71440 (Dec. 21, 1999), corrected, 65 Fed. Reg. 3425 (Jan. 21, 2000).

[47]*Id.* at 1098.

[48]US USPTO, Manual of Patent Examining Procedures § 2107.01, available on the USPTO Web site at *www.uspto.gov/web/offices/pac/mpep* (visited Jan 7, 2005).

[49]66 Fed. Reg. at 1095 (response to Comment 14).

[50]*Id.* at 1096 (response to comment 19). For further discussion of homology-based assertions of utility under the laws of the U.S., the European Union, and Japan, see Trilateral Project B3b, Comparative study on biotechnology patent practices, Theme: Nucleic acid molecule-related inventions whose functions are inferred based on homology search, posted at *www.european-patent-office.org/tws/sr-3-b3b_bio_search.htm.*

Articulations of the utility standard have thus changed over time as the patent system has sought to apply basic principles to newly emerging technologies. Although the latest word from USPTO appears to embrace a more robust standard (and one that is more consistent with Supreme Court precedent) than was suggested by previous formulations, the USPTO guidelines do not have the force of law without judicial endorsement. A case recently decided by a panel of the Federal Circuit, *In re Fisher,* offered an occasion for judicial oversight of the utility standard in the context of claims to nucleic acid molecules. In that case the examiner rejected claims to ESTs encoding fragments of maize proteins for failure to disclose a specific and substantial utility for the claimed molecules. The USPTO Board of Patent Appeals and Interferences affirmed, rejecting as inadequate the asserted utilities of the ESTs for identification and detection of polymorphisms and for use as probes or as a source of primers. On appeal, the applicant's assignee (Monsanto) has argued that USPTO is applying a heightened utility standard to ESTs corresponding to genes of unknown function without statutory authority. On appeal, the Federal Circuit upheld the board's rejection and approved USPTO's guidelines as comporting with the court's interpretation of the utility requirement.[51]

Regardless of the outcome of this particular appeal involving ESTs, new advances in structural genomics and proteomics inevitably will present USPTO and the courts with further unresolved utility issues in the future. Whenever understanding of the functions and uses of structures lags behind the discovery of the structures themselves, the determination of how much information on practical utility is necessary presents a line-drawing problem. The successful interaction between USPTO and the scientific community about where to draw the line for ESTs provides a model for future interactions concerning how to apply patent law to new types of discoveries as science moves forward.

Novelty and Nonobviousness

Perhaps the most basic limitation on access to the patent system is that one may patent only something that is new. What is "new" or "novel" for patent purposes is a function of how patent law defines the content of the "prior art" in § 102 of the Patent Act. The most important sources of prior art are those that are readily accessible to patent examiners—that is, prior patents and printed publications. Other statutory categories of prior art, including technologies that were

[51]Although the panel decision was split 2 to 1, the dissenter, Judge Rader, expressed sympathy for USPTO's effort to find a tool for rejecting modest advances in the face of decisions such as *In re Deuel* that effectively "deprived the Patent Office of the obviousness requirement for genomic inventions."

previously invented, known, or used by others, are less likely to come to the attention of USPTO at the time of examination, but they may be invoked years later by a defendant in an infringement action who challenges the validity of an issued patent.

In order to defeat a patent for lack of novelty, it is necessary to find every element of the claimed invention present in a single disclosure. If it is necessary to go beyond a single disclosure by, for example, combining the disclosures of multiple references in order to find all of the elements, the basis for challenging the patent is not lack of novelty, but rather "obviousness" under § 103 of the Patent Act, and further limitations apply.

The 1952 Patent Act was the first patent statute to impose an explicit requirement of nonobviousness. Because congressional drafters were unhappy with prior judicial efforts to distinguish between patentable and unpatentable results in terms of the nature of the inventive work done by the inventor, they added to their definition of nonobviousness the following sentence: "Patentability shall not be negatived by the manner in which the invention was made." In other words, the obviousness of the inquiry that led to an invention is out of bounds in assessing the obviousness of the resulting invention. The issue is not what the inventor actually did, but whether the invention would have been obvious at the time it was made to a person of ordinary skill. As a result, the obviousness of an invention is determined from the perspective of a person of ordinary skill in light of the prior art at the time the invention was made. The meaning of this standard has been much contested in the patent system. In a formulation that resonated with the courts of an earlier era, the nonobviousness standard distinguishes the unpatentable work of the "ordinary mechanic" from the patentable advances of more insightful "inventors." In other words, the obviousness of the inquiry that led to an invention is out of bounds in assessing the obviousness of the resulting invention; the inquiry must be done in hypothetical, not actual, terms—that is, it must demonstrate not what the inventor actually did but the invention that would be required of a person of ordinary skill.

Many scientists believe that the nonobviousness standard should exclude from patent protection the results of high-throughput DNA sequencing that can be (and have been) performed by modestly competent research technicians in a mechanized discovery process. Thus far, however, this important requirement for patent protection has failed to exclude the results of routine research and development from patent protection in this particular technological context. The reason for this is a pair of decisions from the U.S. Court of Appeals for the Federal Circuit that reversed rejections of patent claims to genes that were cloned using information about the amino acid sequences of the proteins they encoded.[52] How-

[52]*In re Deuel*, 51 F.3d 1552 (Fed. Cir. 1995); *In re Bell*, 991 F.2d 781 (Fed. Cir. 1993).

ever, obviousness is always assessed at the time an invention is made. As science advances, obviousness advances with it. Thus, work that led to patentable discoveries at one time may no longer overcome the nonobviousness hurdle.

In the early days of gene patenting, the process of cloning the gene for a known protein was fraught with uncertainty and required considerable creativity and skill, but as the field progressed, it soon became an increasingly routine matter, albeit one requiring significant technological expertise and financial investment in the initial development phase. Patent examiners accordingly began to reject patent applications claiming genes encoding proteins for which a partial amino acid sequence had previously been disclosed, reasoning that "when the sequence of a protein is placed into the public domain, the gene is also placed into the public domain because of the routine nature of cloning techniques."[53] This analytical approach appeared to be broadly consistent with prior decisions of the Federal Circuit that had stressed the unpredictability of research strategies used to make previous biotechnology products in holding that those product inventions were nonobvious, rather than focusing more narrowly on the predictable character of the products themselves.[54] But when this analytical approach began calling for rejections of claims to genes that were cloned through the use of what had become predictably successful strategies, the Federal Circuit changed course and repeatedly reversed these rejections on appeal. In the case of *In re Deuel*, the court reasoned that, at the time the Deuel invention was made, a novel chemical generally would not have been presumed obvious unless it was structurally similar to a known compound, and proteins are not structurally similar to DNA sequences. That researchers of ordinary skill in the field, equipped with knowledge of the amino acid sequence, could have used known methods to isolate the corresponding native DNA sequence was, in the court's view, "essentially irrelevant to the question whether the specific [DNA] molecules themselves would have been obvious."

In effect, then, the patentability of a newly sequenced DNA molecule in the early 1990s appeared to depend not on whether the teachings of the prior art made this an obvious and readily achieved next step, but on whether the prior art disclosed structurally similar DNA molecules.[55] This approach to the nonobviousness standard is in growing tension with the perceptions of scientific accom-

[53]*Ex parte Deuel*, 1993 Pat. App. LEXIS 22 (Bd. Pat. App. and Interf. 1993).

[54]E.g., *Amgen v. Chugai*; *Hybritech v. Monoclonal Antibodies*.

[55]See Utility Examination Guidelines, *supra*, 66 Fed. Reg. at 1095 ("As the nonobviousness requirement has been interpreted by the U.S. Court of Appeals for the Federal Circuit, whether a claimed DNA molecule would have been obvious depends on whether a molecule having the particular *structure* of the DNA would have been obvious to one of ordinary skill in the art at the time the invention was made." [citing *In re Deuel* and *In re Bell*]).

plishment among scientists. But because it makes it easy for patent applicants to get past the nonobviousness hurdle, they have no incentive to challenge the rule, and after being repeatedly reversed on this point, USPTO seems to have little interest in raising it again, even though advances in the art might now culminate in a different result. As more DNA sequence information becomes available in databases, even the restrictive approach of the Federal Circuit is likely to lead to obviousness rejections, because most newly sequenced genes are likely to be structurally similar to previously disclosed sequences, given conservation of coding regions in genomes.

In chemical patent practice, if a patent application claims a molecule that is structurally similar to another useful molecule that is disclosed in the prior art, the claimed invention may be deemed *prima facie* obvious, and shifting the burden to the applicant to show that the claimed molecule has surprising or superior properties not possessed by the structurally similar prior art.[56] As more DNA sequence information accumulates as prior art in databases, one would expect to see more *prima facie* obviousness rejections for claimed DNA sequences that are structurally similar to previously disclosed DNA sequences. In order to overcome these rejections, applicants must make a showing of surprising properties that will be difficult to do through mere biology *in silico* without further laboratory research to characterize the sequence more fully and to distinguish it from the prior art.

Advances in proteomics have shown that the relationship between DNA sequence and *protein structure* is less predictable than previously might have been supposed. Newly disclosed protein structures might thus have an easier time satisfying the nonobviousness standard than newly disclosed DNA molecules, given the extensive public databases of DNA sequence information.[57] But by the same token, as more proteomic information becomes publicly available, it should become more difficult to establish novelty and nonobviousness for proteins.

Some claiming strategies in proteomics may be vulnerable to novelty and nonobviousness challenges after patents are issued. Although an issued patent enjoys a presumption of validity,[58] this presumption may be overcome by clear and convincing evidence.[59] Some patent claiming strategies in proteomics seek to claim molecules that fit within a binding site on a protein that has been visualized by inference from Cartesian coordinates obtained for the crystallized protein.[60] Such a claim might be drawn to a "pharmacophore" having a specified

[56]*In re Dillon*, 919 F.2d 688 (Fed. Cir. 1990) (en banc).

[57]The patent eligibility and disclosure requirements might still present significant obstacles to patenting proteomics inventions.

[58]35 U.S.C. § 282.

[59]*Connell v. Sears, Roebuck & Co.*, 722 F.2d 1542 (Fed. Cir. 1983); *Boehringer Ingelheim Vetmedica, Inc. v. Schering-Plough Corp.*, 320 F.3d 1339, 1353 (Fed. Cir. 2003).

[60]These claims raise issues concerning the adequacy of disclosure that are addressed in the next section.

spatial arrangement of atoms predicted on the basis of structural information for a receptor and molecular dynamics calculations, or to a compound defined by such a pharmacophore. Such claims could potentially cover a wide range of compounds, but their breadth makes them potentially vulnerable to future prior art challenges. It is difficult to compare these claims to the prior art because prior art compounds typically have not been characterized in a way that makes it apparent in the course of a quick search whether they meet the claim limitations or not. If a product in the prior art meets the limitations of a claim, the claim is invalid even though the product characteristics that are recited in the claim (such as the spatial arrangement of chemical elements) were merely inherent in the prior art product and had never been explicitly described. This is because the discovery of new properties for an old product does not make the old product patentable.[61]

The nonobviousness requirement in U.S. law has as its counterpart in Japanese and European law the concept of "inventive step." The European Patent Convention considers an invention to involve an inventive step if "having regard to the state of the art, it is not obvious to a person skilled in the art." The convention does not define "person skilled in the art." Section 29(2) of the Japanese patent law mandates that a claimed invention will lack an inventive step when that step easily could be made by a person with ordinary skill on the basis of inventions publicly known or worked prior to the filing of the patent application. Participants in this committee's Trilateral Workshop in Bellagio, Italy, which included representatives of the biotechnology sections of the Japanese and European Patent Offices, generally agreed that differences between the United States on the one hand and Europe and Japan on the other in interpretations of these very similar statutory formulations represented the most important difference in policy and practice in the area of genomic- and protein-related patents in the aftermath of the *Deuel* and *Bell* decisions in the United States. In short, these participants characterized "inventive step" as a significantly higher hurdle to obtaining a European or Japanese patent in the field than the interpretation of "nonobviousness" is to obtaining a patent in the United States. This view is supported by some commentators on the trilateral studies of the three offices' approaches to protein structure and EST patents (Shimbo et al., 2004; Howlett and Christie, 2004).

Disclosure

In order to obtain a patent, an inventor must provide in the application a written description of the invention, an enabling disclosure that would allow a person having ordinary skill in the field to make and use the invention without undue experimentation, and a disclosure of the best mode contemplated by the

---

[61]See, e.g., *Titanium Metals Corp. v. Banner*, 778 F.2d 775 (Fed. Cir. 1985).

inventor for making the invention.[62] Patent applicants may not remedy deficiencies in their disclosure obligations after filing their applications without losing the benefit of their original filing dates (and thereby risking loss of rights if intervening prior art has a bearing on patentability).[63] This disclosure becomes freely available to the public upon issuance of the patent or 18 months after filing if a corresponding application is filed anywhere in the world other than in the United States.[64]

The written description requirement has become quite robust in the recent jurisprudence of the Federal Circuit, particularly as applied to genomics inventions. The Federal Circuit has stressed the distinctness of the "written description" requirement from the "enabling disclosure" requirement, holding that it is not enough to provide an enabling disclosure of how to make a product that is not described in the specification.[65] In a series of cases involving claims to DNA sequences, the Federal Circuit has said the "written description" standard, which serves to ensure that the inventor was "in possession" of the invention as of the patent application filing date, requires disclosure of information about the structure of products covered by the claim, not just a description of their function.[66]

---

[62] 35 U.S.C. § 112.

[63] 35 U.S.C. § 132(a).

[64] 35 U.S.C. § 122(b).

[65] *Amgen, Inc. v. Hoechst Marion Roussel, Inc.*, 314 F.3d 1313, 1330 (2003); *Regents of the University of California v. Eli Lilly*, 119 F.3d 1559, 1568 (Fed. Cir. 1997); *Fiers v. Revel*, 984 F.2d 1164, 1170-71 (Fed. Cir. 1993). Judge Rader, joined by Judges Gajarsa and Linn, has argued that the written description requirement should be coextensive with the enabling disclosure requirement except in cases involving priority disputes. *Enzo Biochem, Inc. v. Gen-Probe Inc.*, 42 Fed. Appx. 439, 445 (Fed. Cir. 2002) (dissenting from denial of re hearing en banc). See also *Moba, B.V. v. Diamond Automation, Inc.*, 325 F.3d 1306, (Fed. Cir. 2003) (Rader, J. concurring). *Cf. id.* at (Bryson, J. concurring) (noting that nothing in the language of §112 would justify construing written description and enablement as distinct requirements only in cases involving priority disputes).

[66] *University of California v. Eli Lilly & Co., supra*, 119 F.3d 1559 at 1568 ("In claims to genetic material . . . a generic statement such as 'vertebrate insulin cDNA' or 'mammalian insulin cDNA,' without more, is not an adequate written description of the genus because it does not distinguish the claimed genus from others, except by function.... It does not define any structural features commonly possessed by members of the genus that distinguish them from others. . . . A definition by function . . . does not suffice to define the genus because it is only an indication of what the gene does, rather than what it is."); *Fiers v. Revel, supra*, 984 F.2d at 1171 ("Claiming all DNA's that achieve a result without defining what means will do so is not in compliance with the description requirement; it is an attempt to preempt the future before it has arrived."). *Cf. Enzo Biochem, Inc. v. Gen-Probe Inc.*, 296 F.3d 1316, 1324-25 (Fed. Cir. 2002) (approving of USPTO Guidelines indicating that "functional descriptions" might satisfy the written description requirement "when coupled with a known or disclosed correlation between function and structure"). In addition, in *Capon* v. *Eschhar* (August 2005), Judge Newman held that USPTO had erred in imposing a *per se* rule requiring that nucleic acid sequences be recited in a patent specification when they were known in the field.

Disclosure of an amino acid sequence for a protein and a strategy for cloning the corresponding gene might be enough to satisfy the enablement standard, but it is not enough to satisfy the written description requirement, as elaborated by the Federal Circuit. USPTO has published guidelines and examiner training materials for the written description requirement that explain in some detail its application to genomic inventions,[67] and written description is emerging as a significant constraint on proteomics claiming strategies as well.[68] Although at least one member of the Federal Circuit has called for the court to reconsider decisions expanding the reach of the written description requirement,[69] it appears to retain the support of a majority of the court.

The recent opinions[70] responding to the request for *en banc* reconsideration of the Federal Circuit panel decision in the case of *University of Rochester* v. *G.D. Searle*[71] show the continuing vitality of written description and illustrate its potential to defeat proteomics claims that seek to reach through to future compounds that might be found through the use of protein structure information. The patent at issue in that case arose out of the discovery by scientists at the University of Rochester that stomach irritation associated with nonsteroidal anti-inflammatory drugs is caused by the inhibition of a protective enzyme (PGHS-1 or Cox-1) that is distinct from a similar enzyme (PGHS-2 or Cox-2) that causes inflammation. They hypothesized—correctly, as it turned out—that molecules that selectively inhibit Cox-2 only, without inhibiting Cox-1, might provide relief from pain and inflammation while reducing these gastrointestinal side effects. More recent studies have confirmed that these drugs have serious cardiovascular side effects (Bresalier et al., 2005). Without identifying or testing any such molecules, the University of Rochester obtained a patent on a "method for selectively inhibiting PGHS-2 in a human host, comprising administering a non-steroidal compound that selectively inhibits activity of the PGHS-2 gene product to a human host in need of such treatment," and brought a patent infringement action against pharmaceutical firms that, meanwhile, had developed selective Cox-2 inhibitor products. The district court granted the defendants' motion for summary judgment of patent invalidity for failure to comply with both the written descrip-

[67]USPTO, Guidelines for Examination of Patent Applications Under the 35 U.S.C. 112, ¶1, "Written Description" Requirement. The examiner training materials are available on the web at *www.usUSPTO.gov/web/menu/written.pdf.*

[68]For an analysis of how written description might constrain proteomics claims, see Trilateral Project WM4, Comparative studies in new technologies. Report on comparative study on protein 3-dimensional (3-D) structure related claims (2002).

[69]*Enzo Biochem, Inc. v. Gen-Probe, Inc.*, 296 F.3d 1316 (Rader, J., dissenting from denial of rehearing en banc); *Moba, B.V. v. Diamond Automation, Inc.*, 325 F.3d 1306, (Fed. Cir. 2003) (Rader, J. concurring).

[70]*University of Rochester v. G.D. Searle*, 375 F.3d 1303 (Fed. Cir. 2004).

[71]358 F.3d 916 (Fed. Cir. 2004).

tion and enablement requirements, and the Federal Circuit affirmed on the written description ground without reaching enablement.[72] The university's motion for rehearing *en banc* was denied, but generated a set of concurring and dissenting opinions that revealed significant divisions within the court regarding the proper scope of the written description requirement. Nevertheless, the resolution of this case suggests that written description is likely to pose a significant obstacle to "reach-through" claims to compounds defined functionally in terms of the proteins with which they interact, rather than structurally (as in the past).

Claims to compounds defined in relation to a hypothetical pharmacophore might present a better case for satisfaction of the written description requirement than the purely functional definition set forth in the *University of Rochester* claims. Arguably the disclosure of crystal coordinates and binding sites provides enough structural information about the size and shape of the claimed compounds, linked to the function of binding the target, to permit visualization of the molecules falling within the scope of the claim.[73] But USPTO has indicated that such a claim would fail the written description standard "because one skilled in the art would conclude that the inventors were not in possession of the claimed invention."[74]

For now, it appears that proteomics inventors are likely to be limited to claiming the actual proteins and peptides that they have disclosed and characterized in their patent applications, without being able to reach through to claim as yet unidentified compounds that ultimately might be found to interact with those proteins and peptides.

Limitations on Licensing and Enforcement

License terms for patented inventions pertaining to genomics and proteomics have sometimes been as controversial as the underlying patents themselves. Many biomedical researchers have limited resources to make up-front payments for access to materials that are used mainly in research, such as clones, reagents, and animal models. This has led some providers of these so-called research tools to propose contingent payment terms in the form of reach-through royalties on future product sales or reach-through licenses to future inventions. These terms have the advantage of making tools available at minimal up-front cost for use in noncommercial research, while still permitting the tool owner to share the wealth

[72] *University of Rochester v. G.D. Searle*, 358 F.3d 916 (Fed. Cir. 2004).

[73] See Dep't of Comm., U.S. Pat. & Trademark Off., Guidelines for Examination of Patent Applications Under the 35 U.S.C. 112 ¶1, "Written Description" Requirement, 66 Fed. Reg. 1099, 1106 (Jan. 5, 2001) (stating that written description requirement may be met by disclosure of "functional characteristics when coupled with a known or disclosed correlation between function and structure").

[74] Trilateral Project WM4 at p. 28.

if the research yields a commercial product. But many tool users balk at agreeing to reach-through terms. As such terms become more common in proposed research tool licenses, the obligations imposed by different tool providers may come into conflict[75] or create mounting royalty obligations that reduce incentives for future product development. Moreover, it may be difficult in the future to trace a particular discovery or product to prior use of a research tool and to establish that it is subject to the reach-through obligation. These difficulties increase the transaction costs of negotiating over terms of access to proprietary research tools, slowing their dissemination and delaying research. More generally, reach-through royalties impose a cost on product development, diluting incentives of downstream innovators to reward upstream innovators who may have no continuing involvement in the project. Such an allocation might undermotivate downstream research and development.

Patent Misuse

Some users of research tools have argued that the use of reach-through provisions in licenses should be impermissible under the common law doctrine of "patent misuse,"[76] but so far the courts remain unpersuaded. In *Bayer AG* v. *Housey Pharmaceuticals, Inc.*,[77] a pharmaceutical firm argued that the owner of patents on screening methods to identify potential pharmaceutical products was misusing its patents by licensing them on terms that required the payment of reach-through royalties on future products that were not themselves covered by the patent claims but that were identified through use of the patented screening methods. The district court concluded that, although it would be patent misuse for a patentee to "condition" a license upon the payment of royalties on unpatented products and activities,[78] reach-through royalty terms are nonetheless permissible for the mutual convenience of the parties, when the evidence indicates that the patent holder was willing to consider other payment options.

Judicial deference to agreed license terms generally makes good sense. If the

---

[75]For example, some tool providers may request an option to take an exclusive license to future discoveries made by the user. A researcher who uses more than one proprietary tool may only promise such an option once, and may have already provided such an option to a research sponsor. Even a precommitment to extend a nonexclusive license to use future discoveries would conflict with an obligation to extend an exclusive license to use the same discoveries.

[76]"Patent misuse" means improper exploitation of a patent, e.g., by violating the antitrust laws or by extending the patent beyond its lawful scope. If misuse is found, the patent may not be enforced until the misuse has been purged by abandoning the abusive practice and dissipating any harmful consequences. 6 Chisum on Patents § 19.04 (2000 and Supp. 2003).

[77]228 F. Supp. 2d 467 (2002).

[78]*Bayer AG v. Housey Pharmaceuticals, Inc.*, 169 F. Supp. 328 (2001) (denying patent holder's motion to dismiss claim of patent misuse on the pleadings).

biomedical research community finds it difficult to arrive at mutually agreeable terms of exchange for research tools, the courts should not aggravate the problem by foreclosing options that might help the parties strike a deal. Reach-through provisions could help the bargaining parties resolve disputes about valuation and enable resource-poor institutions to gain access to unaffordable research tools by financing the license through deferred payments that come due if and only if the research yields a successful product.

Experimental Use Exemption

Despite patent law's requirements that the patentee disclose the knowledge underlying an invention and the way in which it is made and utilized, biological information may be used effectively only when researchers can examine, and experiment on, the products and processes subject to the patent. Many scientists believe that their use of patented inventions as the subject of research does not (or at least should not) subject them to infringement liability. The U.S. Patent Act has no general statutory exemptions for noncommercial, experimental, or research uses of an invention, apart from a provision added as part of the Hatch-Waxman Act of 1984 to permit the use of a patented invention "solely for uses reasonably related to the development and submission of information under the Federal law which regulates the manufacture, use, or sale of drugs. . . ."[79] Many other nations provide somewhat broader exemptions. The European Commission's proposed Council Regulation on the Community Patent excludes from the effects of a community patent "acts done privately and for non-commercial purposes" and "acts done for experimental purposes relating to the subject-matter of the patented invention."[80] The national patent laws of many European Union member states contain similar provisions, as does Japanese law.[81]

In a line of cases going back to the 1813 opinion of Justice Story in *Whittemore* v. *Cutter*,[82] the U.S. courts have recognized an "experimental use" defense to patent infringement as a theoretical matter, but they have generally declined to apply the defense to the facts of the cases before them.[83] Justice Story

---

[79]35 U.S. Code § 271(e)(1).

[80]Commission of the European Communities, Proposal for a Council Regulation on the Community Patent, Art. 9 (Aug. 1, 2000), Official Journal of the European Communities C 337 E/278-90 (Nov. 28, 2000), available at http://*www.mdjuris.com/itlaw/ce3372000en.pdf* (visited Dec. 19, 2002).

[81]For a thoughtful critique of U.S. law and comparison to the laws of other nations on this point, see J. Mueller, "No 'Dilettante Affair': Rethinking the Experimental Use Exception to Patent Infringement for Biomedical Research Tools" *Wash. L. Rev.* 76:1-66 (2001).

[82]29 F. Cases 1120 (D. Mass. 1813).

[83]See, e.g., *United States Mitis Co. v. Carnegie Steel Co.*, 89 F. 343 351 (C.C.W.D. Pa. 1898) (noting that defendant's use of patented invention "was a commercial use, extending over a period of

argued that "it could never have been the intention of the legislature to punish a man, who constructed a [patented] machine merely for philosophical experiments, or for the purpose of ascertaining the sufficiency of the machine to produce its described effects."[84] But although subsequent courts have consistently acknowledged that the defense might be available in an appropriate case, only one case, *Ruth* v. *Stearns-Roger Manufacturing Co.*,[85] has generated a published opinion squarely holding that use of a patented invention in a university laboratory qualifies for the defense.

The Federal Circuit has been signaling its discomfort with the experimental use defense for over 20 years. In its first encounter with the defense in the 1984 case of *Roche* v. *Bolar*,[86] the Federal Circuit rejected the argument of a generic drug manufacturer that the defense applied to its use of a patented drug to conduct clinical trials during the patent term. The court characterized the defense as "truly narrow" and refused to extend it to a use that was "no dilettante affair such as Justice Story envisioned" but rather had "definite, cognizable, and not insubstantial commercial purposes." Shortly thereafter, Congress amended the patent statute as part of the Hatch-Waxman Act[87] to provide a defense from infringement liability for the use of a patented invention "solely for uses reasonably related to the development and submission of information under a Federal law which regulates the manufacture, use, or sale of drugs. . . ."[88]

In *Madey* v. *Duke University*,[89] the Federal Circuit rejected the common law experimental use defense as applied to academic research, declaring the noncommercial character of the research irrelevant to its analysis of the case. What matters to the Federal Circuit is whether the research "is in keeping with the alleged infringer's legitimate business, regardless of commercial implications." In the case of a major research university, noncommercial research projects "unmistakably further the institution's legitimate business objectives, including educating and enlightening students and faculty participating in these projects." Activities

several months, and involved a very large product"); *Bonsack Machine Co. v. Underwood*, 73 F. 206, 211 (C.C.E.D.N.C. 1896) (noting that machine "has not been made simply as an experiment, but has been used for profit, that is, for the purpose of selling the [defendant's] patent"); *Albright v. Celluloid Harness-Trimming Co.*, 1 F.cas. 320, 323 (1877)(No. 147) (holding use of patented invention in trial manufacture "is a technical infringement, and is sufficient to authorize an injunction restraining . . . future use" but not sufficient for award of damages); *Poppenhusen v. Falke*, 19 F. Cas. 1048, 1049 (C.C.S.D.N.Y. 1861)(No. 11,279) (noting that defendants "are rivals of the complainant in the very business to which his patents relate").

[84]29 F. Cas. at 1121.

[85]13 F. Supp. 697 (D. Colo. 1935).

[86]733 F.2d 858 (Fed. Cir. 1984).

[87]35 U.S.C. § 271(e).

[88]This defense is codified at 35 U.S.C. §271(e)(1).

[89]307 F.3d 1351 (Oct. 3, 2002).

that further these "business objectives," including research projects that "increase the status of the institution and lure lucrative research grants, students and faculty," are *ipso facto* ineligible for the experimental use defense.

In contrast to the restrictive position of the Federal Circuit toward the scope of the common law research exemption, the recent decision of the U.S. Supreme Court in *Integra* v. *Merck*[90] took an expansive approach toward the scope of the Hatch-Waxman statutory research exemption. The alleged infringement in that case involved the use of patented peptides to assess their potential therapeutic efficacy in the inhibition of angiogenesis in the course of a collaboration between scientists at the Scripps Research Institute and Merck KGaA. Although Merck KGaA had originally raised both the common law experimental use defense and the Hatch-Waxman statutory exemption in the district court, after the *Madey* decision came down it decided to focus its appeal exclusively on the statutory exemption, arguing that its use of the peptides was reasonably related to the development of information for submission to FDA. On appeal, the Federal Circuit panel majority held that the statutory exemption did not extend to the sort of preclinical research at issue in that case. The majority opinion addressed the common law experimental use exemption only in a footnote, noting that the common law exemption was not before the court, although questioning whether the exemption would be available even if the issue had been properly raised.[91] In a dissenting opinion, however, Judge Newman argued that the Scripps/Merck activities should have qualified either for the common law exemption or the statutory exemption.

The Supreme Court granted review to consider the scope of the statutory exemption and reversed the Federal Circuit by a unanimous vote.[92] Noting that "the statutory text makes clear that it provides a wide berth for the use of patented drugs in activities related to the federal regulatory process," the court rejected the argument that the statutory exemption covers only clinical trials and not preclinical research.[93] On the other hand, the court recognized some limits to the scope of the exemption, noting that "[b]asic scientific research on a particular compound, performed without the intent to develop a particular drug or a reasonable belief that the compound will cause the sort of physiological effect the research

---

[90]331 F.3d 860 (Fed. Cir. 2003), *cert. granted,* 160 L. Ed. 2d 609 (2005).

[91]*Id.* at note 2 ("[T]he common law experimental use exception is not before the court in the instant case. . . . [T]he Patent Act does not include the word 'experimental,' let alone an experimental use exemption from infringement. . . . [T]he judge-made doctrine is rooted in the notions of *de minimis* infringement better addressed by limited damages.")

[92]*Merck KGaA v. Integra Lifesciences*, 2005 U.S. Lexis 4840 (2005), bench opinion available at *http://a257.g.akamaitech.net/7/257/2422/13jun20051230/www.supremecourtus.gov/opinions/04pdf/03-1237.pdf* (visited June 14, 2005).

[93]*Id.* at 8-9.

intends to induce, is surely not 'reasonably related to the development and submission of information' to the FDA."[94] Although some *amicus curiae* briefs urged the court to consider the common law exemption as well as the statutory exemption, [95] it did not do so. The court's decision thus broadens the scope of the statutory exemption to cover some preclinical research activities in the course of drug development, without offering any relief from infringement liability for basic research at an earlier stage before drug development and FDA submissions are dominating the course of research and development. On the other hand, the Court's opinion made clear that even failed attempts at development may be considered part of the efforts to generate research data for FDA submission.[96]

Some observers believe that *Merck* v. *Integra* leaves open a number of important questions. First, it states that "[b]asic scientific research on a particular compound, performed without the intent to develop a particular drug or a reasonable belief that the compound will cause the sort of physiological effect the researcher intends to induce, is surely not 'reasonably related to the development and submission of information' to the FDA."[97] It provides no guidance, however, on how to implement the line drawing envisioned. Second, because the court refers to patented compounds, and not to patented inventions, it raises questions about whether patented processes or other patented inventions that are used in the course of drug development but are not the intended subject of an FDA submission should be treated in the same way. This ambiguity is intensified by the court's statement that it "need not—and do[es] not—express a view about whether, or to what extent, [the statutory exemption] exempts from infringement the use of 'research tools' in the development of information for the regulatory process."[98]

Finally, there is comparable precedent in federal statute. The 1996 Ganske-Frist amendment to the infringement statute of U.S. patent law (35 USC § 287) exempts from infringement liability medical practitioners who perform patented medical or surgical procedures that do not employ a patented device or process, so long as the procedure is carried out in association with a health care entity such as a medical clinic, university, or hospital. Currently, the amendment specifically excludes biotechnology patents or any patent tied to molecular biological methods and life science.

---

[94]*Id.* at 12.

[95]See, e.g., Amicus Brief of the Bar Association of the District of Columbia, Brief of the Consumer Project and the EFF as Amicus Curiae in Support of Merck KGaA. These and other briefs in the case are posted on the Internet at *http://patentlaw.typepad.com/patent/2005/02/merck_kgaa_stat.html* (visited May 9, 2005).

[96]2005 U.S. Lexis at 13-14.

[97]*Merck KGaA v. Integra Lifesciences*. 125 S.Ct. 2372, 2382 (2005).

[98]*Id.* at 2382, note 7.

Compulsory Licensing

The United States has consistently taken the position that there should be no general provision for the compulsory licensing of patents.[99] However, lawmakers in other countries think differently. Several nations provide by law for a compulsory license in the event that the patentee refuses to engage in the activities required to fully disseminate the invention. For example, in the United Kingdom, the state is permitted to compel licensing if a patented invention is not commercialized "to the fullest extent that is reasonably practical" for three years.[100] German patent law provides for a compulsory license if the patentee is not willing to grant a license to someone who offers "reasonable compensation."[101] Although there has been little litigation under this provision in recent years, German practitioners claim it has a significant *in terrorem* effect on patentees. Other examples include the Competition Act, § 32 (Canada); the Patent Law 1999, § 93 (Japan); and the Patents Act 1990 (Cth), §§ 133, 163-167 (Australia).

Even in the United States, patents are sometimes subject to government control. Royalty-free licenses are used occasionally as a remedy in anti-trust litigation.[102] Compulsory licenses have been created by statute when important national interests are at stake.[103] Furthermore, the federal government and its contractors can use inventions without authorization, but subject to the payment of just compensation.[104] The federal government also enjoys so-called march-in rights to inventions created with government funding. When the rights holder fails to take steps to commercialize such an invention within a reasonable time, the federal agency that provided funding may step in and grant licenses to parties that are willing to bring the invention into public use.[105] Accordingly, even the general sentiment against compulsory licensing should be understood as giving way when important public interests are at stake.

The Agreement on Trade-Related Aspects of Intellectual Property Rights also permits compulsory licensing in response to particular problems. It permits members to use compulsory licenses to control anti-competitive practices.[106] Furthermore, it allows members to permit unauthorized uses when specified con-

---

[99]See, e.g., *Dawson Chemical v. Rohm & Haas*, 448 U.S. 176 (1980) (noting "the long-settled view that the essence of a patent grant is the right to exclude others from profiting by the patented invention" and that "[c]ompulsory licensing is a rarity in our patent system".)

[100]Patents Act, 1977, c. 37 § 48.

[101]German Patent Act § 24 (Law of December 16, 1980, as amended Dec. 9, 1986).

[102]See, for example, *United States* v. *General Electric Co.*115 F.Supp. 835 (D.N.J.1953). See also *Charles Pfizer & Co. v. Federal Trade Comm'n*, 401 F.2d 574 (6th Cir.1968) (requiring licensing at a reasonable royalty), cert. denied, 394 U.S. 920, 89 S.Ct. 1195, 22 L.Ed.2d 453 (1969).

[103]See, e.g., the Clean Air Act, codified at 42 U.S.C. §7608.

[104]28 U.S.C. 1498.

[105]35 U.S.C. § 203.

[106]TRIPS Agreement, art. 40.

ditions are met.[107] Each case must be considered on its individual merits; use must be limited in time and manner to the purposes authorized; any license granted must be on a nonexclusive and nonassignable basis, use generally must be limited to local needs; and the rights holder must be paid adequate remuneration. Except in national emergencies, the rights holder must be given the opportunity to license on its own and, in every case, the rights holder must be notified as soon as is reasonably practical.

Patent Pooling

A patent pool is an agreement between two or more patent owners to license one or more of their patents to one another or to third parties.[108] In general, pool members assign or exclusively license their intellectual property to a separately administered entity, which then controls the licensing of the patent portfolio back to the members and, if the pool is "open," to third parties as well. Members might be entitled to use the bundle of intellectual property on a royalty-free basis, or they might have to pay. In addition, licensing revenue generated by the pool can be divided in several different ways, and management of the pool might involve diverse voting structures or veto rights.

Patent pools remove intellectual property barriers to the exploration of technology, promote the integration of complementary technologies, and reduce the transaction costs of obtaining multiple licenses. They also are viewed sometimes as a cheaper and faster way to resolve some disputes than litigation would be (Clark et al., 2000; OECD, 2005). Patent pools have been identified as another possible solution to any eventual biotechnological anti-commons.

The potential role of patent pools in the biotechnology industry has received considerable attention in the literature. Some analysts have argued that in view of possible royalty stacking, anti-commons, and other situations in which existing patent rights could become impediments to further research and development, patent pools have significant benefits and therefore should be encouraged. For example, a 2000 white paper issued by USPTO promoted their use, stating:

> The use of patent pools in the biotechnology field could serve the interests of both the public and private industry, a win-win situation. The public would be served by having ready access with streamlined licensing conditions to a greater amount of proprietary subject matter. Patent holders would be served by greater access to licenses of proprietary subject matter of other patent holders, the generation of affordable pre-packaged patent "stacks" that could be easily licensed, and an additional revenue source for inventions that might not otherwise be developed. The end result is that patent pools, especially in the bio-

[107]TRIPS Agreement, art. 31.

[108]See Klein, *supra at www.usdoj.gov/atr/public/speeches/1123.html.*

technology area, can provide for greater innovation, parallel research and development, removal of patent bottlenecks, and faster product development. (USPTO, 2000, p. 11)

Patent pools, like most licensing arrangements, usually are beneficial to competition. They may, however, occasionally reduce or eliminate it. Nevertheless, patent pools are uncommon in the biotechnology industry so far (OECD, 2005). Concerns exist about whether patent pools could solve problems in markets for genetic inventions. For example, a recent Organisation for Economic Co-operation and Development survey of biotechnology and pharmaceutical companies, genetic testing centers, and public research organizations found that respondents did not consider patent pools or cross-licensing agreements to be helpful in increasing access to genetic inventions (OECD, 2002). Respondents cited the difficulty of measuring the value of the contributions that each party would bring to the arrangement. Even though the utility of patent pools for the biotechnology industry is arguable, it generally is accepted that patent pools may benefit both patent owners and consumers, provided the pool is limited to complementary and/or blocking patents. Under these conditions, patent pools may facilitate the integration of complementary technologies, reduce transaction costs, facilitate the clearing of blocking patent positions, and avoid infringement litigation (Bratic et al., 2005; OECD, 2005).

Some features particular to the biopharmaceutical industry may create disincentives to patent pooling. Patent pools are more common in industries that use patents defensively, whereas in the biopharmaceutical industry patents tend to be used offensively. Patent holders who enter a pool risk losing the more significant revenue they might receive if they exclusively licensed their patents. Biotechnology companies generally prioritize the accumulation of patents, which makes them attractive for buy-out, while large pharmaceutical companies have tended to buy the intellectual property and/or small companies they need. However, large companies no longer have the capital to continue buying whatever intellectual property they need, which may create an incentive for more patent pooling. There also may be an incentive for pooling when the complexity of intellectual property requires a pool for research to progress; one example is the international patent pool for SARS. In order to avoid antitrust liability, all patents in the pool must be essential for the particular aim. A pool that is enabling is probably justified—the existence of multiple licenses to the pool creates a safe harbor —while a pool that restricts output or use is more problematic. There also is cause for concern if the pool revolves around price setting or standard setting, although this concern is often addressed by the federal antitrust agencies' exercising oversight of a pool's operation.

CONCLUSIONS

Although the same patent laws apply to all fields of technology, new technologies inevitably present USPTO and the courts with new problems in the interpretation and application of old standards to determine such issues as patent eligibility, utility, novelty, nonobviousness, and adequacy of disclosure. Because the resolution of legal disputes takes time, a lag between the emergence of new technologies and the resolution of disputed issues of patent law that the new technologies raise will occur. Most of the existing legal precedents involving genomic patents address technology that is at least a decade old. Important issues concerning the patentability of ESTs—a technology from the early 1990s—are only now being addressed by the Federal Circuit. Meanwhile, USPTO and the scientific community have had notable success in working together to determine how best to apply the standards of patent law to new types of discoveries in genomics, providing a model for addressing emerging patent law questions in proteomics and structural genomics in a more timely fashion.

4

Trends in the Patenting and Licensing of Genomic and Protein Inventions and Their Impact on Biomedical Research

This chapter reports the committee's findings with respect to its charge to determine current trends in the patenting of genomic- and protein-related inventions, licensing practices, and the impact on biomedical research and innovation. To address these issues to the greatest extent possible within the limits of the available time and resources, the committee consulted the existing research literature and received testimony from scholars in various fields, government officials, and stakeholders. In addition, the committee engaged in three original research efforts:

1. a search for issued patents and published patent applications in selected biotechnology categories;
2. a small survey of university licensing of selected categories of patents. This and the first effort supplemented information being gathered systematically by other investigators in larger-scale research studies; and
3. a survey of biomedical research scientists to ascertain their experience with intellectual property and its effects on research.

The first and second tasks were performed by committee staff. The third, more ambitious project was a survey of approximately 2,000 randomly selected researchers in universities, industry, and government laboratories. It was conducted by John Walsh and Charlene Cho, University of Illinois at Chicago, and Wesley Cohen, Duke University, and it was supported by funding from the committee.[1] It builds on a more limited interview-based survey by Walsh, Cohen, and Ashish Arora—work carried out for the National Academies' predecessor Committee on Intellectual Property Rights in the Knowledge-Based Economy (NRC, 2003). The new survey represents the first systematic effort to shed light on the concerns expressed by members of the academic community that patents on upstream discoveries may impede follow-on research and development if access to the foundational intellectual property is restricted or is too difficult, time consuming, or costly to obtain. The new survey goes further, however, to try to determine the extent of biomedical researchers' involvement with intellectual property, its role—positive as well as negative—in decisions to initiate, redirect, or suspend research, and investigators' experience with sharing of research data and materials, whether or not protected by intellectual property. The survey achieved a modest response rate and is subject to the limitations of an inquiry relying on memory and self-reporting, but its results are largely consistent with the findings of the earlier nonrandom interviews. The results of these inquiries and the committee's interpretation of those results and of closely related studies are presented in this chapter.

TRENDS IN PATENTING GENOMIC AND PROTEIN INVENTIONS

Although not the only source of data on genomic and proteomic patents,[2] the most extensive database of U.S. "gene" patents was initiated by the congressional Office of Technology Assessment in the early 1990s with assistance from the United States Patent and Trademark Office (USPTO) and Georgetown University scholars and was transferred to Georgetown University, where it is maintained and continually updated with the support of the National Institutes of Health (NIH) and the Department of Energy. Using a proprietary patent database, Delphion, the investigators have compiled a comprehensive set of patents from several broad biology-related patent classes. These are patents that refer to nucleic acids and closely related terms assembled into an algorithm to search in their claims. From 1971 until 2006, approximately 33,000 issued nucleic acid patents have been identified. The annual rate of patenting did not exceed 500, however,

---

[1]The full report, J. Walsh, C. Cho, and W. Cohen, *Patents, Material Transfers, and Access to Research Inputs in Biomedical Research*, June 2005, is available at *http://www.uic.edu/~jwalsh/NASreport.html*.

[2]See also A.M. Michelsohn, *Biotechnology Innovation Report 2004: Benchmarks* and *Biotechnology Innovation Report.* Washington, DC: Finnegan, Henderson, Farabow, Garett & Dunner, LLP. These sources have reported numbers and ownership of patents in several biotechnology categories, identified by key word searches. The results are not incompatible with those described below, but the use of carefully delineated search algorithms yields more discriminating results than do keyword searches.

until 1991 when it began accelerating, peaking at 4,500 in 2001. The number of issued patents declined sharply in 2002 and again in 2003 and 2004 (Figure 4-1).

A more refined analysis has been done using bioinformatics methods to compare nucleotide sequences claimed in U.S. patents to the human genome (Jensen and Murray, 2005). This analysis shows that approximately 20 percent of human genes (4,382 of the 23,685 genes currently in the public databank) are explicitly claimed, not merely disclosed, in issued U.S. patents owned by 1,156 different assignees. A number of genes, including BRCA1, are subjects of multiple patents asserting rights to various gene uses and manifestations. In a few cases, single patents claim multiple genes, usually as probes on a DNA microarray. None of these circumstances is by itself indicative of a "thicket" or "blocking" problem absent information on patent claims, licensing, and corporate relationships.

Large numbers of applications for patents with such claims are still pending, some of them since the early 1990s. Many of these can be retrieved from the database, because USPTO began publication of most 18-month-old applications in March 2001; but there are two reasons why the precise number for each year cannot be ascertained. First, under the American Inventors Protection Act of 1999, an application can be withheld from publication if the filer agrees not to seek a patent on her or his invention outside the United States. In biotechnology and organic chemistry the "withholding" rate was about 6 percent through 2002 (NRC, 2004). Second and more important, USPTO has not been systematic about publishing applications after 18 months of filing. For example, applications filed in 2001 and 2002 continue to appear for the first time in the database in 2005.[3]

Approximately 5,000, or 15 percent, of the issued patents in the Georgetown database are managed by universities, led by the University of California, the largest patent holder in the field overall. The dominance of the University of California is somewhat misleading, however, because the number is a compilation of the patenting activity of 10 major research institutions, including the University of California at San Francisco, the University of California at Berkeley, and the University of California at San Diego. More than 800 are assigned to the U.S. government. The government also has an "interest" in as many as 60 percent of the patents held by the leading academic patenting institutions, meaning that they derived from federally funded research.[4] The majority of patents are held by U.S.-based biotechnology and pharmaceutical companies. Figure 4-2 shows the 30 largest holders of DNA-based U.S. patents.

[3]The Georgetown University investigators observed this when recently adding pending DNA patent applications to their database.

[4]A federal grantee is supposed to disclose in a patent application the government's "interest" in the invention, but it is doubtful that this rule is followed consistently. Thus, the 60 percent figure for university genomic inventions may in fact be on the low side.

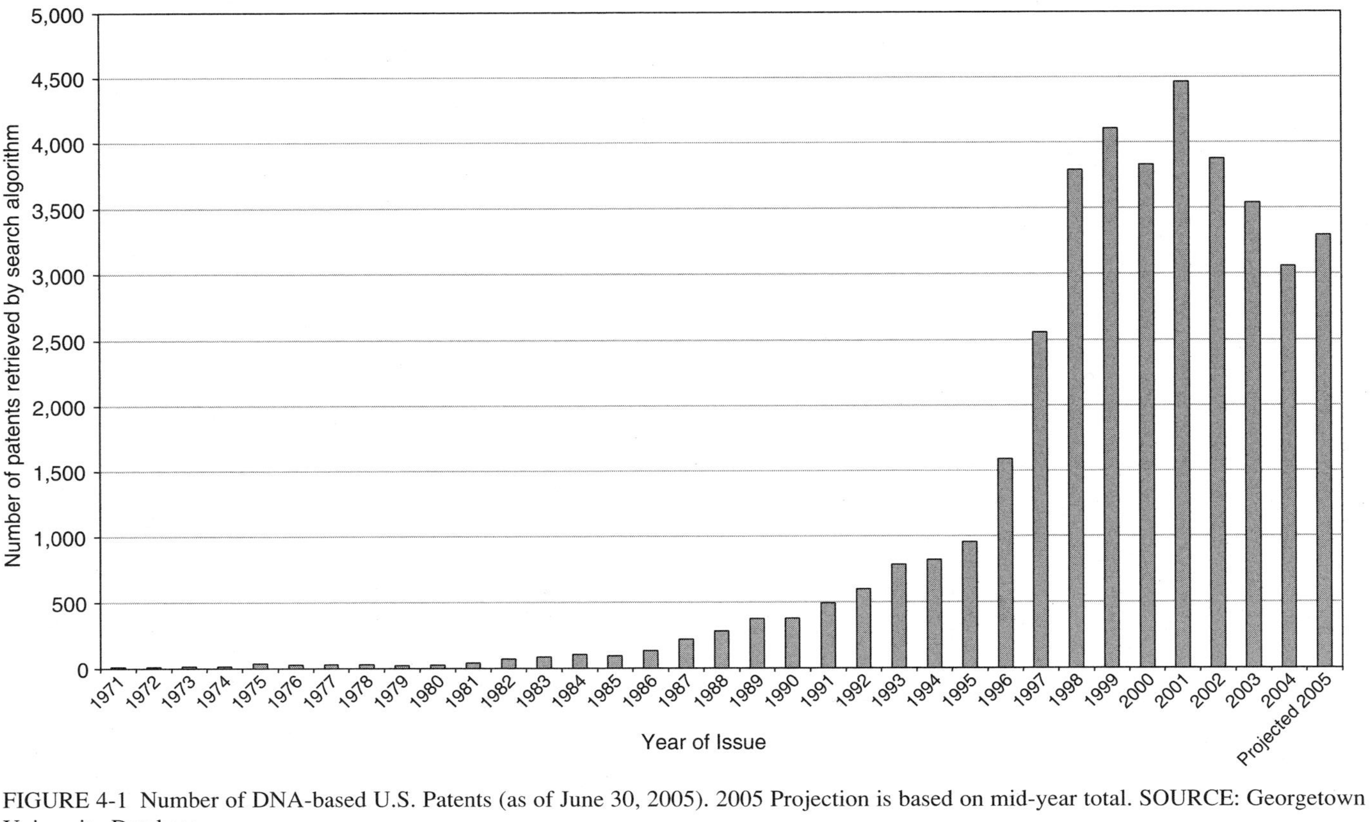

FIGURE 4-1 Number of DNA-based U.S. Patents (as of June 30, 2005). 2005 Projection is based on mid-year total. SOURCE: Georgetown University Database.

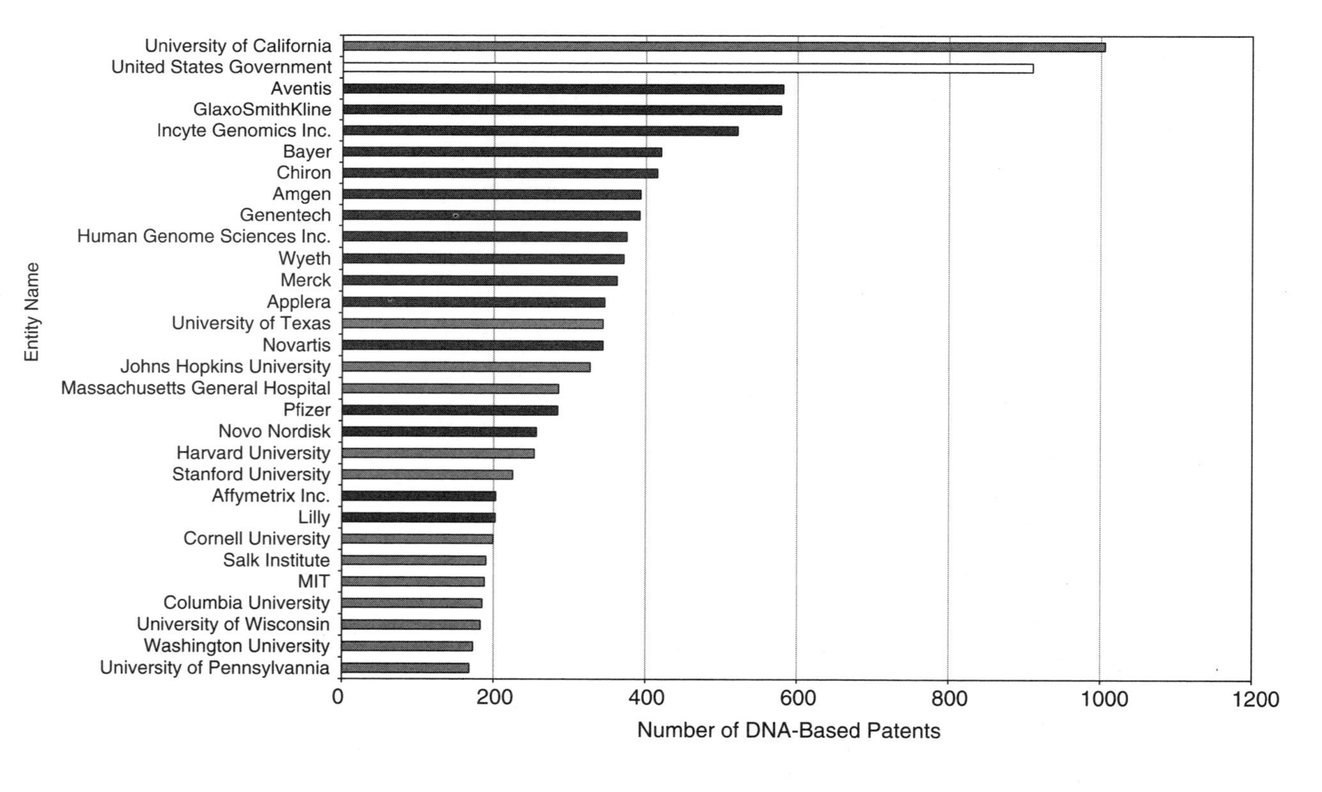

FIGURE 4-2 Thirty entities holding the largest number of DNA-based U.S. patents (as of June 30, 2005). SOURCE: Georgetown University database.

In his charge to the committee at its first meeting, Francis Collins, Director, National Human Genome Research Institute, requested data on what patents have been issued or applied for, by whom, and in which countries for nine more specific categories of genomic- and protein-related patents: gene regulatory sequences, single nucleotide polymorphisms (SNPs) and haplotypes, protein structures, protein-protein interactions, gene expression profiling, genetically modified organisms, and related software, algorithms, and databases.

Further discussions among the committee resulted in the selection of three additional patent categories, each representing a distinct disease-related molecular pathway: Cytotoxic T-Lymphocyte Associated Protein-4 (CTLA4), Epidermal Growth Factor (EGF), and Nuclear factor-Kappa B (NF-kB).[5] CTLA4, EGF, and NF-kB were chosen from a longer list of known pathways on the basis of four criteria:

1. they are seen as involved in or correlated with more than one category of disease, spanning cancer and autoimmune or inflammatory diseases;
2. there is significant scientific research interest, as indicated by frequent citation in the scientific literature;
3. they exhibit some variance in the number of related patents; and
4. there is some but varying industry involvement, represented by pharmaceutical or biotechnology firm patenting activity, licensing of university patents, or clinical testing or even marketing of therapeutic products.

---

[5]The following description is based on Walsh, et al., 2005. The biological pathways regulated by EGF, CTLA-4, and NF-kB are recognized widely by the biomedical research community for their roles in mediating disease and normal development. Stimulation of cells with EGF, for example, has been shown to induce cell division, an event that, if left unchecked, can lead to cancerous growth. The CTLA-4 receptor is involved in regulating T cell proliferation, and its loss of function is believed to contribute to autoimmune diseases such as rheumatoid arthritis, multiple sclerosis, and lupus. NF-kB also has been implicated in rheumatoid arthritis as well as asthma, septic shock, and cancer, and its role in the proper development and function of the immune system is supported by numerous studies of NF-kB knockout and transgenic mice.

The intense interest of the scientific community in these pathways is reflected in scientific publications and in the patenting of composition of matter products and/or processes related to EGF, CTLA-4, and NF-kB. Foundational papers on EGFand NF-kB each have been cited more than 1,500 times, while the more recent discovery of the functions of CTLA-4 has yielded more than 900 citations. Since 1995, USPTO has granted more than 760 EGF-related patents, 90 NF-kB patents, and more than 60 CTLA-4 patents distributed among large pharmaceutical firms, biotechnology firms, universities, and the federal government.

There are also on the market or in development several therapeutic products targeted to these proteins. For example, both Erbitux® (ImClone/Bristol-Myers Squibb) and Iressa® (AstraZeneca) are used for the treatment of cancers associated with EGF receptor expression. CTLA4-Ig® (Repligen) and Abatacept® (Bristol-Myers Squibb) also are patented and currently are in clinical trials for the treatment of multiple sclerosis and rheumatoid arthritis, respectively.

In short, it is reasonable to hypothesize that to the extent they arise at all, intellectual property complications will be greater in research involving at least some of these pathways than in genomic and proteomic research in general.

Methods

In consultation with USPTO supervising examiners in technology center "1600" (biotechnology), committee staff developed search algorithms for each of the categories of patents (see Appendix C). These searches were run on the patent claims field to obtain the number of U.S. patents and assignees, assignee countries, inventor countries, application years, and ultimate assignees over the period from January 1, 1995, to February 1, 2005. An independent search using the same algorithms for the same period was made subsequently by staff of the Georgetown University project. The numbers of patents found in the two sets of searches corresponded very closely but not exactly. In addition to U.S.-assigned patents, the searches included published U.S. patent applications and, for comparison, patents and applications issued by the European Patent Office (EPO). The "software," "database," and "algorithm" categories were limited to patent classification 435 (chemistry: molecular biology and microbiology). Table 4-1 summarizes the results. Especially for the "software" and "algorithm" categories, the class restriction may limit the results, because biologically related patents may have been placed in other patent classifications. It was not possible with the re-

TABLE 4-1 Issued U.S. and European Patents and Patent Applications in Selected Categories of Biotechnology Inventions, 1995-2005

Category	U.S. Granted	U.S. Application	EPO Granted	EPO Application
Genes and gene regulation	6,145	7,105	1,327	1,153
Haplotype/SNPs	1,482	2,292	266	293
Gene expression profiling	7,428	16,983	2,635	3,043
Protein structure	39	230	28	31
Protein-protein interactions	6,964	12,845	3,590	2,066
Modified animals	652	2,767	177	334
Software	60	209	11	28
Algorithms	91	325	64	113
Databases	1,466	3,765	A	A
EGF pathway	765	1,045	212	166
CTLA4 pathway	63	149	19	19
NF-kB pathway	94	206	42	81

NOTE: A = No biological class restriction is available for this category.

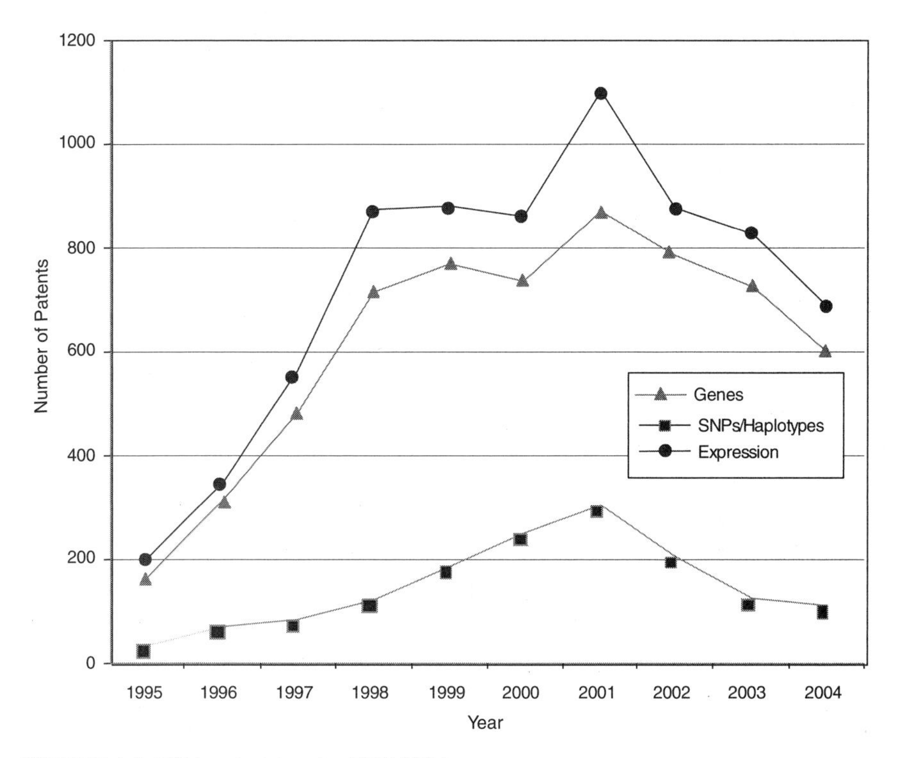

FIGURE 4-3 DNA patent trends, 1995-2004.

sources available to restrict the searches to human material, excluding plants, animals, microorganisms, and synthetic molecules.[6]

The patenting trends and published applications by year from 1995 to 2004 are shown in Figures 4-3 through 4-7, with the nine categories grouped as follows: DNA patents (including genes and gene regulation, haplotypes and SNPs, and gene expression profiling) and tools (modified animals, software, algorithms, and databases). Protein structures and protein-protein interactions are shown separately because of the vast difference in patenting activity, which is characteristic of other categories. Genes and gene regulation, gene expression profiling, and protein-protein interactions are by far the most active categories, followed by haplotypes and SNPs and databases. There are few protein structure patents and

[6]This will be apparent in Table 4-2, in which some agricultural biotechnology firms appear as leading patent holders in some categories, especially genes and gene regulatory sequences and SNPs and haplotypes.

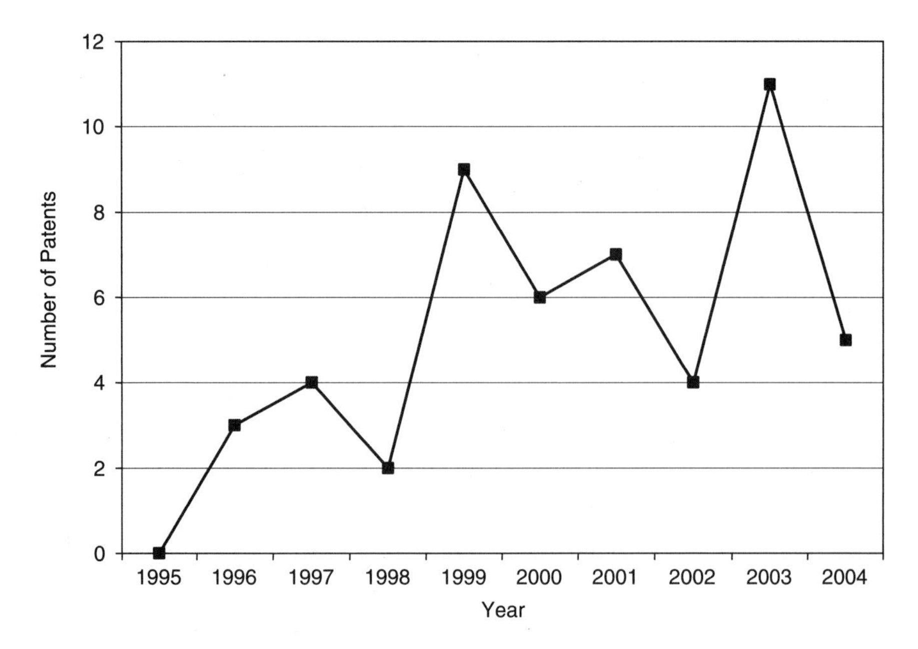

FIGURE 4-4 Protein structure patent trends, 1995-2004.

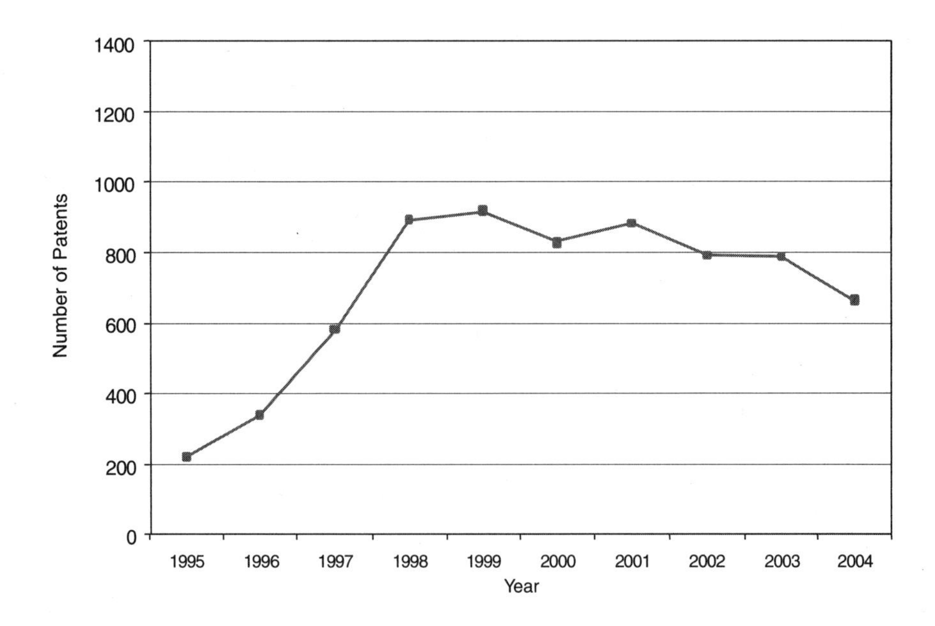

FIGURE 4-5 Protein-protein interactions patent trends, 1995-2004.

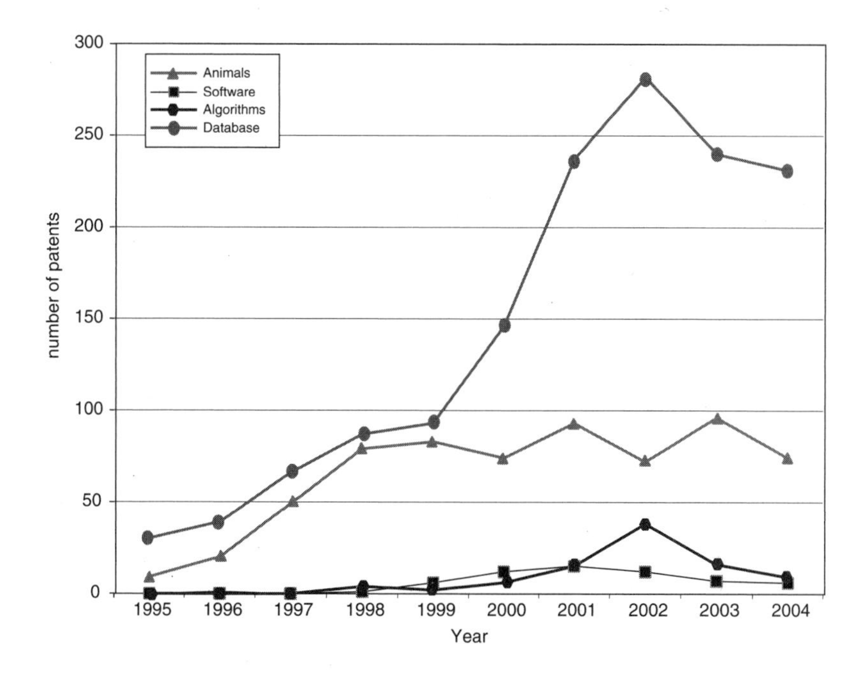

FIGURE 4-6 Research tools patent trends, 1995-2004.

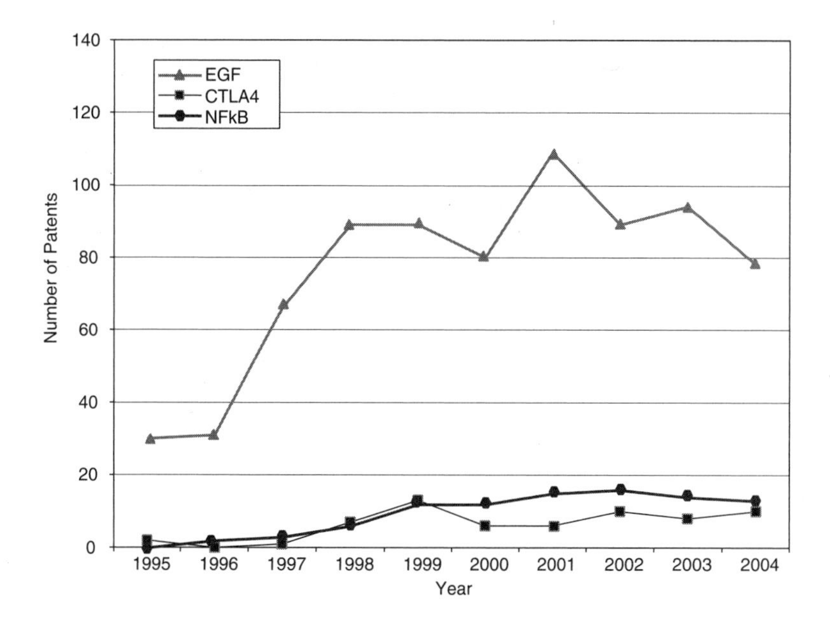

FIGURE 4-7 Molecular pathway patent trends, 1995-2004.

pending applications, as well as few biologically related software and algorithm patents. There are also sharp differences among the pathways. Indeed, that was a criterion of selection. The area of EGF shows considerably more activity than those of CTLA4 or NF-kB. The EPO data show lower levels of patenting in every category but similar variations among categories. Compared to the United States, the low European levels of patenting of haplotypes and SNPs and genetically modified animals are particularly striking and perhaps attributable to greater conservatism on the part of EPO in approving patents in those domains.

Similar to the Georgetown University DNA patent data, patenting declined in most categories beginning in 2000-2001. The only case in which this is not readily apparent is protein structures, where the numbers are very low to begin with. Does this signify a general decline in biotechnology patenting that can be expected to continue? It is of course too early to tell. However, several possible explanations can be ruled out or considered unlikely: (1) public research funding was not declining during this period; in fact, the decline begins at a time when the NIH budget was being doubled; (2) research productivity was not declining; if anything it was increasing, with the automation of sequencing and improvements in other techniques; and (3) the economic environment could not have played a role, at least initially, because the patents that issued after 2000 derived from applications filed two or more years earlier, at the height of the boom.

Greater conservatism on the part of USPTO is almost certainly a factor in the decline, perhaps especially in categories such as haplotypes and SNPs. Partly in response to criticisms of the standards being applied to genomic patent applications, the office conducted a broad review of its examination standards and practices and in January 2001 released new guidelines clarifying the written description and utility requirements. The guidelines are written to be generic to all technologies, but they had a significant effect on claims involving DNA and proteins, and most of the training examples given to examiners are in biotechnology. The written description guidelines were intended to bring USPTO practice into line with the Federal Circuit's decision in *Regents of the University of California* v. *Eli Lilly and Co.*,[7] in which the court ruled that simply describing a method for isolating a gene or other sequence of DNA is insufficient to show possession and that the complete sequence or other identifying features must be disclosed. The guidelines declared that the claimed utility of the invention must be "specific, substantial, and credible" and extend beyond a mere description of its biological activity. The guidelines were widely interpreted as raising the bar to patents on genomic inventions (see Chapter 3).

[7]*Regents of the Univ. of Cal.* v. *Eli Lilly & Co.*, 119 F.3d 1559 available at U.S. App. LEXIS 18221, 43 U.S.P.Q.2d (BNA) 1398 (Fed. Cir. 1997).

But the question of what practical effect the measures had on examiners' behavior and USPTO output is difficult to answer. It is complicated by the lag between application filings and patent grants and other nearly simultaneous developments, such as the deposit of large amounts of human DNA sequence data in the public domain, where they became prior art. Other factors to be weighed in interpreting patent grants over time are the finite nature of the human genome and USPTO's "restriction" practice of forcing patent applicants to separate DNA sequences into different applications.

Markets may have had another kind of influence. The rising cost of patenting may have discouraged some from vigorously pursuing all of the patenting opportunities presented by the flourishing of biomedical research.[8] Moreover, there is evidence that technology licensing offices become more sophisticated and selective as they accumulate experience about what technologies are licensable (Mowery et al., 2004).

Another change can be documented and may be the principal effect of the new examination policies—the lengthening of patent application pendency. By the end of 2002, applications in the principal category affected by the guidelines (526/23.1, DNA/RNA fragments, including ESTs) were taking more than 3 years to yield patents, several months longer than the 24-month average in 2002 for all applications (Figure 4-8).

The data indicate that there are at least as many applications pending in USPTO as there are patents already issued in each of the patent categories, and in some cases—for example, gene expression profiling, protein-protein interactions, and modified animals—two to four times as many.[9] In EPO, the numbers of patents and applications are less divergent, but in most categories more applications are pending than patents have been issued.

No one can predict how many of the pending applications eventually will issue, let alone how many will be filed in the future. The grant rate in USPTO (the proportion of applications that result in issued patents) is substantial—two-thirds to 90 percent or more by various calculations, and probably higher than the approval rates in the European and Japanese patent offices.[10] The committee concluded that the patent landscape, which already is crowded in areas such as gene expression and protein-protein interactions, could become considerably more complex over time.

---

[8]Pressman et al. (2005) report that a number of university technology transfer officials share this view.

[9]For the reasons mentioned above, not all applications older than 18 months have been published.

[10]Because of the practice of continuation applications—refilings with modifications but retaining the original filing or priority dates—calculation of patent grant or allowance rates in the United States is complex and the methodologies controversial (National Research Council, *A Patent System for the 21st Century*, 2004). They have not been applied to particular technological areas.

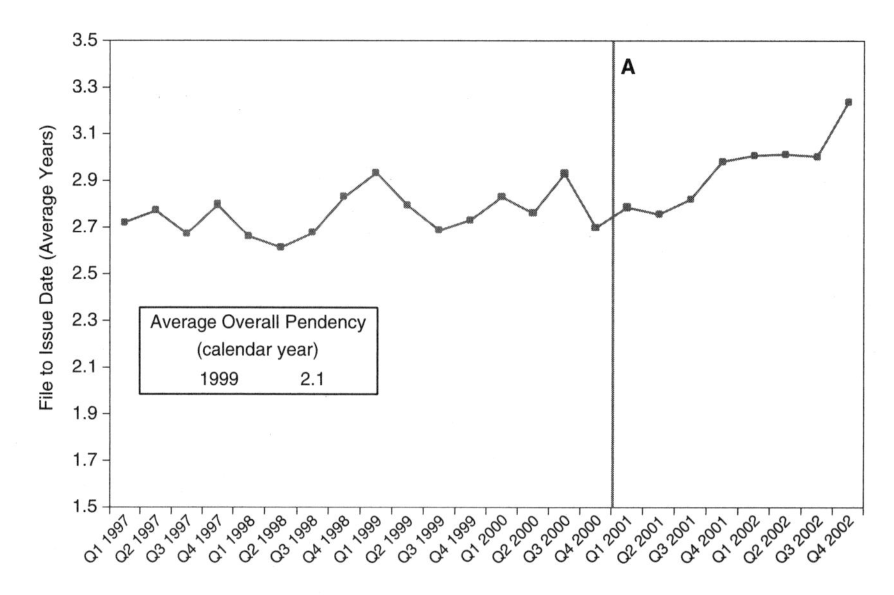

FIGURE 4-8 Patent Class 536/23.1 (DNA/RNA fragments) application pendency by quarter, 1997-2002. NOTE: Pendency is calculated based on original file date and issue date for all issued patents in Class 536/23.1. Overall pendency is calculated by USPTO and also includes an estimate of the time from filing date to abandonment of the application. A = Utility and written description guidelines implemented.

The committee staff analysis also looked at the inventor and assignee country for the patents in each of the categories. The United States leads the world in both inventors and assignees, holding 65 to 80 percent of the worldwide patent share, followed by the United Kingdom, France, Germany, Japan, and Canada. Other active countries are Israel and the Netherlands. For the U.S.-granted patents and pending U.S. applications, the top assignees were all in the United States—sometimes by a factor of 15 or more. Additionally, the United States is the top inventor country, but by a smaller margin. Figures 4-9 through 4-12 show these data for the DNA-related categories, the protein-related categories, tools, and the three pathways.

In the United States, the institutions constituting the University of California have been active patentees across all categories, with a combined patenting frequency at the level of a commercial entity. Categories that developed more recently (software, databases, and gene expression profiling) typically are dominated by biotechnology firms. The protein structure category has been led by pharmaceutical companies because of its proximity to drug discovery, while universities have been dominant in the modified animals category. NIH (through the

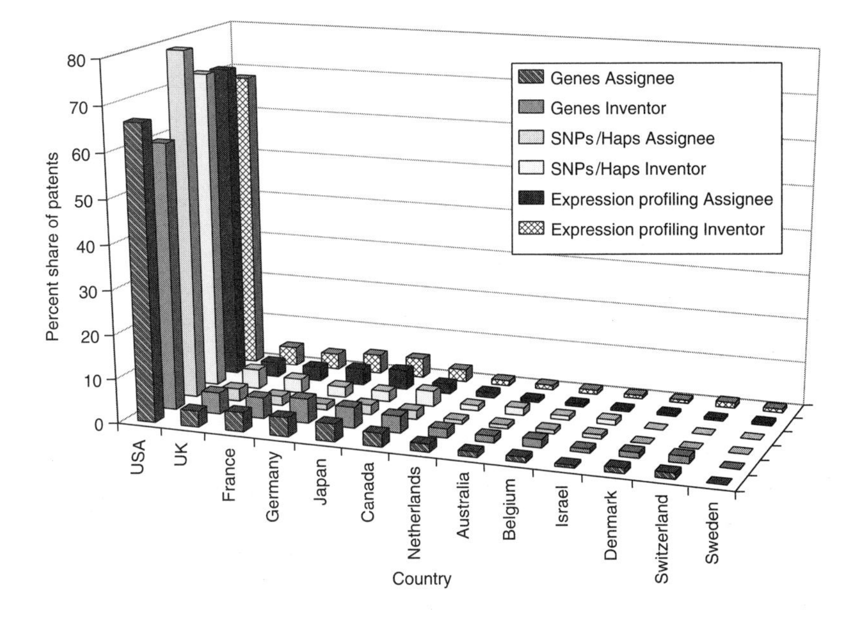

FIGURE 4-9 DNA patents: inventor and assignee country, 1995-2004.

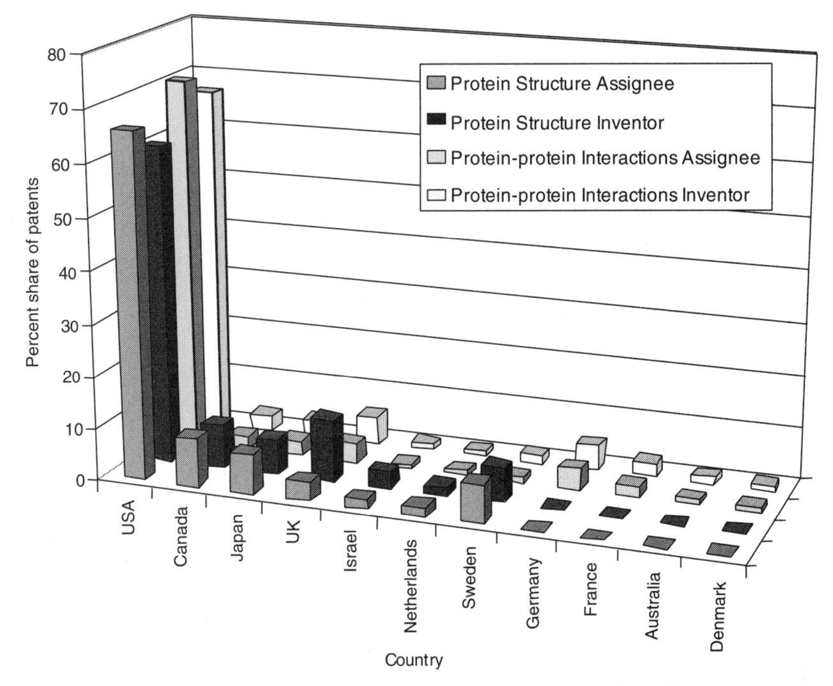

FIGURE 4-10 Protein patents: inventor and assignee country, 1995-2004.

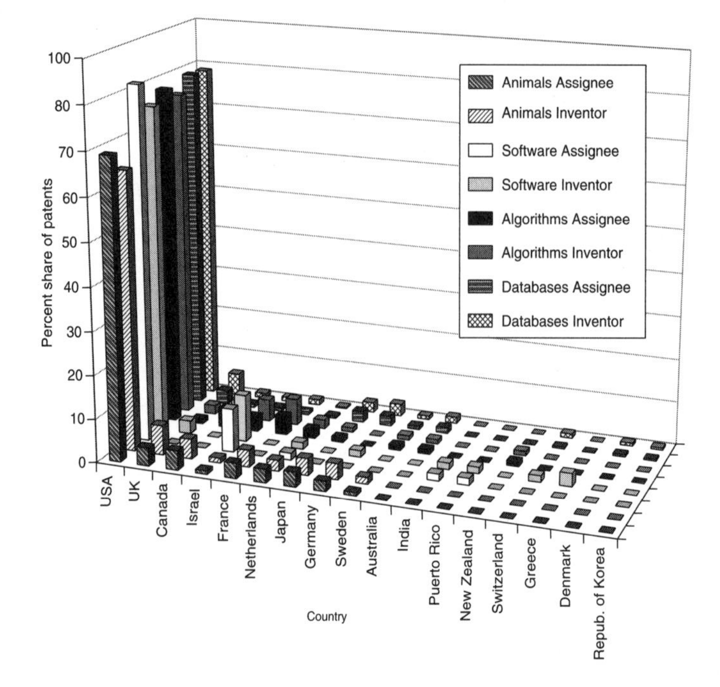

FIGURE 4-11 Research tool patents: inventor and assignee country, 1995-2004.

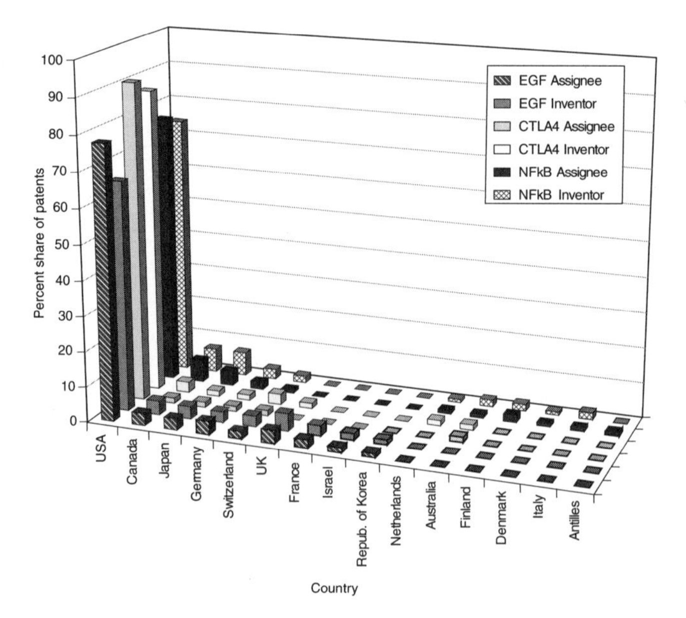

FIGURE 4-12 Molecular pathway patents: inventor and assignee country 1995-2004.

Department of Health and Human Services) also has been active in several categories, especially gene-expression profiling and protein-protein interactions (Table 4-2). Table 4-2 shows the leading assignees of patents in the 12 categories described above and identified using the search algorithms set out in Appendix C.

TABLE 4-2 Principal Assignees of Patents by Category

	Total Patents	Top Assignees
Genes and Gene Regulatory Sequences	6,145	*U. California (188)*
		Pioneer Inc. (150)
		Ludwig Inst. (72)
		Monsanto (72)
		Chiron Corp. (71)
		General Hosp. (71)
SNPs and Haplotypes	1,482	Pioneer (183)
		Dekalb Genetics (107)
		Stine Seed Farm (48)
		U. California (39)
		John Hopkins (25)
Gene Expression Profiling	7,428	*U. California (215)*
		Incyte (170)
		Affymetrix (117)
		Gen-Probe (100)
		DHHS (96)
Protein Structure	39	Abbott Labs (3)
		Connaught Labs (3)
		U. California (3)
		U. Alberta (3)
Protein-Protein Interactions	6,964	Genentech (181)
		U. California (178)
		DHHS (84)
		Chiron (82)
		Immunex (78)
Modified Animals	652	*U. California (26)*
		General Hosp. (11)
		Pharming BV (10)
		Abgenix Inc. (9)
Software	60	Millennium (8)
		Rosetta (4)
		Pioneer Hi-Bred (3)

continued

TABLE 4-2 Continued

	Total Patents	Top Assignees
Algorithms	91	Cytokinetics (42) All others (2 or 1)
Databases	1,466	Affymetrix (108) *U. California (45)* Agilent Tech. (34) Nanogen (22) Sequenom (18)
EGF	765	Sugen (23) Genentech (16) *U. California (12)* **DHHS (12)** *Yale (11)*
CTLA4	63	Bristol-Myers Squibb (20) *Dana Farber (6)* Repligen (4) Genetics Inst. Inc. (3) Pfizer (3)
NF-kB	94	*U. California (7)* Bristol-Myers Squibb (6) Tularik (5) Ariad (3) *Dalhousie Univ. (3)*

NOTE: The assignee is the company or organization assigned ownership on the original patent. Through consolidations, mergers and acquisitions, and other transactions, ownership may change. Private organizations, foundations, and hospitals are distinguished from commercial entities by *italics*. Government entities are indicated by **bold** typeface.

TRENDS IN UNIVERSITY LICENSING OF GENOMIC AND PROTEOMIC INVENTIONS

Licensing is the principal means of accessing the use of patented technology, and it occurs under terms that are infinitely varied and complex and whose effects are not straightforward. Thus, whether a patented upstream invention is available for licensing—and under what conditions—is possibly the principal determinant of whether exclusive rights can impede or alternatively, facilitate the conduct of follow-on research. Unfortunately, data on licensing are very limited for two reasons. First, unlike patents, government administrative processes generate information on only some licenses whose representativeness is unknown and for which detailed information may be lacking and not publicly available. Licenses to or by publicly traded companies are required to be reported to the U.S. Securities and Exchange Commission, but only if they are "material" to the financial performance of the firms. Research grantees are required to report invention disclosures, patents, and licenses to technologies developed with federal support to a repository—the Edison database, which is maintained by NIH—but these data were not accessible by the committee. Generally speaking, not only firms but also universities consider licenses and licensing terms proprietary information that they voluntarily disclose very selectively and only when it is to their advantage.

Despite these sensitivities, some progress has been made in identifying the parameters of licenses of university-owned biotechnology patents.[11] In particular, the Georgetown University group surveyed 30 U.S. academic institutions owning 75 or more of the DNA-based patents in their database (which have been described previously) at the beginning of 2003 (Pressman et al., 2005). Nineteen institutions provided data on licensing frequency for about 2,700 patents. Detailed data were obtained on 200 licensing agreements involving 500 patents, supplemented by qualitative responses to open-ended policy questions and phone interviews.

The principal findings were as follows:

- Approximately 70 percent of the patents managed by survey respondents were licensed at least once, in a large majority of cases before the patents were issued; approximately 2 percent were licensed more than 9 times.
- For patents licensed once, 56 percent were licensed exclusively for all fields, 36 percent were licensed exclusively by field of use, and only 8 percent were nonexclusive. For patents licensed 2 to 9 times, 36 percent were nonexclusive, 46 percent were exclusive by field, 13 percent were exclusive for all fields (often sequentially), and 5 percent were licensed on other terms. The types of licenses used vary a great deal by company type, with all-field exclusive licensing dominant in the case of start-up enterprises but less frequent with established companies regardless of size (Figure 4-13).
- Nevertheless, the results underscore the fact that exclusivity is not a reliable indicator of the extent to which patented inventions are available for others to use.[12] Some patents are licensed exclusively but to multiple entities for many different fields of use. Exclusive licenses terminate for a variety of reasons, and patents are subsequently relicensed. In some cases, licenses are renegotiated and their exclusivity modified, and some exclusive licenses permit sublicensing with or without the agreement of the patent holder.

---

[11]Licensing data are collected annually from university technology transfer offices and reported by the Association of University Technology Managers, but they are not broken down by field of technology or research.

[12]Nor is it highly correlated with commercialization (Thursby et al., 2005).

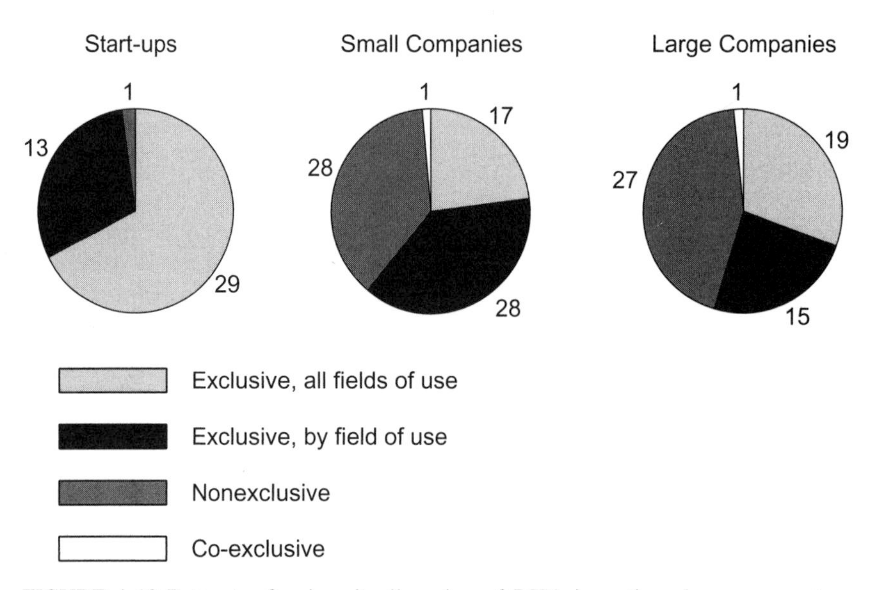

FIGURE 4-13 Patterns of university licensing of DNA inventions by company type. SOURCE: Pressman et al., 2005.

• Qualitative responses from the universities suggest that the utility and development potential associated with a technology have an important influence on both patenting and licensing behavior. When known utility and the presumed potential for commercial development are low, universities are less inclined to patent, and when they do, licensing tends to be nonexclusive. On the other hand, when utility and presumed commercial potential are both high, universities are inclined not only to patent but also to license exclusively.

• Most exclusive licenses contain nonfinancial due-diligence requirements, as do about 45 percent of nonexclusive licenses. These are requirements to report progress in further development of the technology and steps in commercialization. Among the 62 responding universities, 78 percent said that they had terminated research because of due-diligence problems.

• Most institutions report reserving rights to use a patented technology for their own investigators even though it is licensed exclusively to a commercial entity. An increasingly common university practice in recent years is to reserve such rights for investigators at other nonprofit institutions, but this is often subject to the patent holder's case-by-case approval.

In short, interview respondents reported practices broadly consistent with the NIH Research Tool guidelines issued in 1998 and with the *Guidelines for Licensing of Genomic Inventions*, which were in draft form and published for comment at the time the survey was conducted and the results analyzed. Further, university technology transfer offices reported considering the NIH guidelines *de facto* regulations binding on grantee institutions.

This committee also obtained the cooperation of the five university assignees with the most patents (in all but one case) holding patents on inventions related to the three molecular pathways—CTLA4, EGF, and NF-kB—to supply data on the licensing of these inventions. The results were similar to those obtained by the Georgetown team for DNA-based patents held by an overlapping set of research institutions. Eleven institutions reported on a total of 122 patents—86 EGF patents, 17 CTLA4 patents, and 19 NF-kB patents. Approximately two-thirds of the patents have been licensed at one time or another—75 once and 7 two or more times. The remaining patents have been abandoned or their licensing histories are unknown. Of 90 licenses reported, two-thirds (60) are exclusive for all fields, 21 percent are exclusive by field of use, and 12 percent (11) are nonexclusive. Most of the agreements (90 percent) allowed sublicensing. Approximately 60 percent were licensed to start-up firms (overwhelmingly on exclusive terms), about 21 percent to firms identified as biotechnology companies, and 20 percent to pharmaceutical firms. Diligence requirements of some sort are included in nearly all license agreements involving these pathway patents.

EFFECTS OF INTELLECTUAL PROPERTY PRACTICES ON RESEARCH[13]

The implications of patenting and licensing practices are likely to vary from one stage of research and development to another—for example, basic, curiosity-driven research; drug discovery and development; clinical and diagnostic testing—and depend on a variety of circumstances, including the resources of the respective parties and their awareness of the existence and use of intellectual property. A patent on an upstream discovery may encourage downstream development if it gives a developer necessary protection from free riding by others. A patent on an upstream discovery may be an impediment to downstream research if it results in lack of access by downstream researchers not in need of exclusivity or to a foundational discovery or indispensable research tool (a "blocking" problem), or if it renders access to multiple patented technologies excessively difficult or costly (the "thicket" or "anti-commons" problem). Efficient licensing practices can help lower transactions costs and reduce these problems of access to key upstream technologies.

To collect more extensive and systematic but still preliminary information on these relationships, the committee arranged with Walsh and colleagues to de-

---

[13]Except where otherwise noted, the empirical findings of this section are drawn from Walsh et al., 2005.

sign and conduct a post-mail survey of biomedical researchers in academia, industry, government, and other nonprofit institutions. The committee also provided comments on a draft survey instrument and the proposed sampling methodology, as did other experts consulted. The sample was drawn from the membership lists of relevant professional societies.[14] Excluded were researchers in academic institutions not among the top 70 recipients of NIH research awards. Investigators identified as working in government or industry automatically were included in the sampling frame because of their under-representation in the lists relative to university investigators. In fact, industry researchers were oversampled to ensure that they constituted about one-third of the total. To ensure that the survey respondents contained sufficient numbers of individuals who work in fields of biomedical sciences of high commercial interest (because of their association with normal and diseased cellular processes), a specially selected sample of approximately 100 researchers working on each of the three molecular pathways described earlier (EGF, CTLA4, and Nf-kB) also was included.

The total sample of 1,125 included investigators in universities, government laboratories, and other nonprofit institutions; 563 industry scientists; and 299 researchers working on one of the signaling proteins. In all, 655 responses were received (a 33 percent unadjusted response rate[15]), 414 from "academia," including government and nonprofit sectors,[16] and 144 from industry. The pathway samples yielded about 30 respondents each.[17]

In keeping with the committee's interests, the Walsh et al. survey asked re-

[14]American Society of Cell Biology, Genetics Society of America, American Crystallographers Association, American Society for Biochemistry and Molecular Biology, American Society for Pharmacology and Experimental Therapeutics, American Association of Immunologists, Biophysical Society, Protein Society, American Society for Clinical Investigation, American Society of Human Genetics, and American Peptide Society.

[15]In view of the modest response rate, respondents and nonrespondents were compared on several variables for which the survey team had measures for both groups—papers published (from PubMed), patents (from the USPTO database), institutional affiliation, and highest degree (MD or PhD). Respondents and nonrespondents had similar institutional affiliations and numbers of publications and patents. There was some difference with respect to highest degree; respondents were more likely to have PhDs and nonrespondents more likely to have MD degrees. In comparing MDs with other respondents on the key questions in the survey, it was found that MDs are less likely (0 percent versus 6 percent) to regularly check for patents, although they are just as likely to say they might need patented technology (7 percent versus 8 percent). There is no difference in being delayed (2 percent versus 1 percent) or stopped (0 percent for both) by third party patents. MDs are more likely to report not having their last request for a research material transfer fulfilled (41 percent versus 15 percent), which suggests that the estimate given below of the incidence of being denied access to others' research materials may be conservative. On the other hand, MDs are not more likely to claim that they were stopped due to negotiations over an MTA (6 percent versus 9 percent).

[16]Sixty-nine percent of "academics" work in universities, 11 percent in hospitals (including university hospitals), and 19 percent in government or other nonprofit institution laboratories.

[17]See the full report for demographic and other descriptive data on the respondents.

spondents to identify their fields, their research objectives, and the size of their research groups in order to distinguish between the fields of genomics (mapping and sequencing of genes and researching gene functions and associations with diseases) and proteomics (determining protein structures, interactions, and cell signaling), and across the various stages of research and development and projects of different scale (e.g., large projects might be expected to use larger numbers of patented research elements). About 40 percent of the academic respondents reported doing genomics research, and just under 40 percent were doing proteomics research. Ten percent indicated that they were doing drug development, clinical research, or developing diagnostic tests, and almost 80 percent said they were performing basic research, with the remainder developing research tools or doing other work. About 70 percent are associated with research groups of 3 to 10 people; 20 percent were 1- or 2-investigator projects; and just fewer than 10 percent were groups of more than 10 people.

Before trying to ascertain the effects of patents on research, the 159-item questionnaire inquired about researchers' involvement in commercial activity and in the generation of intellectual property and about their awareness of other intellectual property bearing on their work. Twenty-seven percent of academic respondents have some research tie with small or medium-sized enterprises, and 16 percent have ties with large firms. Nineteen percent of academic respondents receive funding from industry (an average of 4 percent of their research budgets), a slight decline from 5 years ago when 23 percent reported receiving industry funding.[18] The average academic respondent spends about 3 percent of her or his time on commercial activity, defined as paid consulting, negotiation of intellectual property rights, or working with a start-up based on the researcher's own invention.

Forty-three percent of academic respondents have applied for a patent at some point in their research careers, with about 22 percent having done so in the last two years. They averaged 0.37 percent patents each in the last two years. Thirty percent of academics have been involved in negotiations over the rights to their inventions; 11 percent had begun developing a business plan or laying other groundwork for starting a firm; 8 percent had a start-up based on their invention; 13 percent had a product or process on the market; and 18 percent had some licensing income, with 5 percent of this group receiving more than $50,000 in total. Not surprisingly, for those academics conducting drug discovery, clinical testing, or diagnostics, industry funding and patenting rates are higher. In general, there is much more business activity and there is more licensing income in particular than for those engaged in basic research. For academic investigators work-

---

[18]This result is consistent with the recent stability or slight decline in industry funding of academic research overall (NSF, 2004). The current industry contribution of 7 percent is higher than it was in the 1960s and 1970s but lower than it was in the 1950s.

ing on one of the three molecular pathways, commercial activity was somewhat higher than the average for academics, especially for those involved with NF-kB and EGF, but less so for those working on CTLA4.

The rather high level of commercial involvement, including patenting, contrasts with the rather low awareness of the existence of relevant, already-existing intellectual property bearing on investigators' work, despite the proliferation of patents on elements of upstream research. Only 8 percent of academic respondents (32) indicated that their research over the previous 2 years involved information or knowledge covered by someone else's patent. Nineteen percent reported not knowing, and the other 73 percent expressed confidence that they did not need access to other intellectual property. But do academic biomedical scientists attempt to find out if there are patents impinging on their research? Only 5 percent of respondents do so on a regular basis. The percentage is about twice as high for investigators engaged in drug discovery and research involving NF-kB, but not for those working on other pathways.

In the aftermath of the *Madey* v. *Duke* decision, both institutional concerns and patent asserters are raising awareness somewhat. Approximately 22 percent of academic respondents have been notified by their institutions to be careful with respect to patents on research inputs, up from 15 percent five years ago. Five percent have been notified at one time or another that their own research may be infringing upon another's intellectual property. Those external influences are having only a very modest effect on behavior, however. In the two years since the *Madey* v. *Duke* decision, only 2 percent of academic bench scientists have begun to check regularly for patents that might impinge on their research. Cautionary notifications from institutions are seemingly ineffectual: 5.9 percent of those who report receiving such notices regularly check for patents, compared with 4.5 of those who recall no such advice to consider the intellectual property rights of others.

Does patenting provide a positive incentive for academic investigators to conduct certain kinds of research, apart from the reputational rewards, competitive influences, and norms that govern the behavior of the scientific community?

Although motivations are exceedingly difficult to disentangle, it appears that the patentability of results is not a negligible factor in academic research choices—only 7 percent consider it more than moderately important—but it pales in comparison to scientific importance (97 percent), personal interest (95 percent), feasibility (88 percent), and access to funding (80 percent) as reasons to do the work. Of course, patentability and commercial potential rank much higher (19 percent and 22 percent, respectively) for those conducting research on drugs and other therapies than for the average academic scientist engaged in basic research (4 percent and 6 percent, respectively). Furthermore, intellectual property prospects may have a bearing on the availability of research funding, especially from industry.

To what extent do patents negatively impinge upon research by leading aca-

demic investigators to abandon lines of work they otherwise might pursue or to modify research protocols or by raising costs or causing delays?

To probe the adverse impact of patents on research, the survey questionnaire asked respondents to ". . . think about the most recent case where you seriously considered initiating a major research project and decided not to pursue it at that time" and to rank responses on a 1 ("not at all important") to 5 ("very important") scale. Table 4-3 shows the percentage of academic respondents in each research category and in the random sample as a whole scoring a given reason more than a "3," or more than moderately important.[19]

The reasons for project abandonment were, in order of frequency, lack of funding, conflict with other priorities, a judgment that the project was not feasible, not scientifically important, or not that interesting, and the perception that the field was too crowded with competing investigators. Technology access issues—"unreasonable" terms for obtaining research inputs (10 percent) or too many patents covering needed research inputs (3 percent)—are less frequently cited as important factors. Terms of access weigh more heavily on investigators involved in work on drugs and therapies than on basic researchers (21 percent versus 9 percent), on researchers working on NF-kB than on those involved with other pathways (19 percent versus 7 to 9 percent), on those involved in genomics than on those in proteomics, and on those involved in industry-funded research or other commercial activity than on those who are not.

It is important to follow the experience of the academic respondents, although few in number (32), who concluded that they needed a research input covered by someone else's patents. Twenty-four contacted the patent owner to obtain permission to using the patented input; five proceeded without contacting the owner; and one modified a project to avoid the input. None abandoned the work as a consequence of either delay or inability to receive permission. Of those who sought permission, seven reported not receiving it within one month. A higher proportion of those intending to use the patented technology as a drug experienced delays or difficulty than those intending to use it as a research tool. Of those seeking permission, only one encountered a demand for licensing fees, in the range of $1 to $100.

Overall, the number of projects abandoned or delayed as a result of technology access difficulties is extremely small, as is the number of occasions in which investigators revise their protocols to avoid intellectual property issues or pay high costs to obtain one. Thus, it appears that for the time being, access to patents or information inputs into biomedical research rarely imposes a significant burden for academic biomedical researchers.

---

[19]Without an index allowing comparison across an individual's answers to all questions, the percentages in Table 4-3 do not reflect relative importance. That is, the data do not correct for the fact that some individuals may answer that everything is a 3 or higher.

TABLE 4-3 Reasons for Not Pursuing Projects, by Research Goal and Pathway

		Research Goal			Pathways		
	Random Sample	Drug Discovery	Basic Research	Other	CTLA4	EGF	NF-kB
No Funding	62	86	60	58	63	54	82
Too Busy	60	55	60	59	53	58	48
Not Feasible	46	41	46	47	33	55	53
Not Scientifically Important	40	24	41	45	40	36	50
Not Interesting	35	24	36	33	20	30	29
Too Much Competition	29	21	32	21	27	29	29
Little Social Benefit	15	21	14	15	13	5	22
Unreasonable Terms	10	21	9	6	7	9	19
Not Help w/ Promotion/Job	10	21	7	15	0	13	5
Too Many Patents	3	3	2	3	0	4	0
New Firm Unlikely	3	3	2	3	0	4	0
Little Commercial Potential	2	3	2	3	0	4	0
Little Income Potential	1	3	1	3	0	4	0
Not Patentable	1	3	1	3	0	4	0
Respondents	274	28	213	33	16	24	22

SOURCE: Walsh et al., 2005.

There are, however, reasons to be concerned about the future in addition to the earlier observation that the patent landscape is becoming more complex in many domains of research. First, the lack of substantial evidence for a patent thicket or a patent-blocking problem is associated with the general lack of awareness or concern among investigators about existing intellectual property.[20] That could change dramatically and possibly even abruptly in two circumstances. Institutions, aware that they currently enjoy no legal protection, may become more concerned about their potential patent infringement liability and take more active steps to raise researchers' awareness or even to try to regulate their behavior. The latter could be both burdensome on research *and* largely ineffective because of researchers' autonomy and their ignorance, or at best uncertainty, about what intellectual property applies in what circumstances. It is much easier for corporations to exercise due diligence in the context of research that is centralized and directed than it is for universities, where research is highly decentralized and decisionmaking is fragmented.

On the other hand, patent holders, equally aware that universities are not shielded from liability by a research exception, could take more active steps to assert their patents. The latter may not extend to more patent suits against universities—indeed, established companies may be reluctant to pursue litigation against research universities—but it could involve more demands for licensing fees, grant-back rights, and other terms that raise transaction costs that are burdensome to research. More assertions would, in all likelihood, prompt more defensive behavior on the part of institutions that traditionally are risk averse. Whether proactively in planned research or defensively in response to claims of infringement, established companies typically go to great lengths and considerable expense to determine what constitutes a "valid" patent. If necessary, the in-house legal department will consult outside counsel to verify its views. The resources necessary to conduct patent literature searches and arrive at validity judgments on a frequent or routine basis probably are beyond the capacity of most nonprofit research institutions and a wasteful diversion, in any case. Nevertheless, failure to perform due diligence could limit research institutions' ability to approach demands for licenses by distinguishing between patents that probably are valid and patents that likely would be held invalid in litigation.

According to information collected from 66 research universities by the American Association for the Advancement of Science,[21] there was an increase in patent infringement notifications received in the aftermath of the *Madey* deci-

---

[20]The two conditions likely reinforce each other. The absence of thicket or blocking problems encourages ignorance or inattentiveness and vice versa.

[21]The survey was sent to 240 institutions, with a low response rate of just over 25 percent. It was conducted in association with the Association of American Medical Colleges, the Association of American Universities, the National Association of State Universities and Land-Grant Colleges, and the Council on Government Relations.

sion. Most of them involved biomedical research, although it is not known how many instances pertained to genomic and proteomic patents. The number reported increased from 16 in the January-June 2003 6-month period to 36 in the July-December period. In about half of the cases, the notice was in the form of a request to take a license. These led to background investigations in only four cases and licensing agreements in only about one-fifth of the cases. At the same time, only 9 of the 66 institutions reported having a written policy encouraging faculty to consider whether they might be infringing on intellectual property rights in planning and conducting research.[22] Although so far little disruption of research has occurred, few precautions have been taken to limit the infringement liability exposure of universities. A few cases of successful patent assertions could have a powerful demonstration effect and upset this equilibrium.

The second source of concern is that biomedical research is becoming more complex and increasingly requires larger-scale efforts. The pattern of a single investigator working on a single gene or gene sequence is giving way to more multi-investigator projects entailing work on many genes or proteins simultaneously, more and more of them patented. The Walsh et al. survey, the sample for which included research teams of significant size, did not indicate that intellectual property-related complications are greater in proportion to the number of investigators involved in the effort, but it is a reasonable presumption that such would be the case with more research inputs. Of course, the resources to address intellectual property complexities also are likely to be greater the more substantial the project. Even if the status quo continues—with many investigators and research institutions not taking precautions to avoid infringement and not subject to frequent patent assertions—the absence of any shield from infringement liability raises a further concern. Institutions may encounter difficulties in licensing the inventions of their researchers in the future.

Are the effects of intellectual property on research different for work on the three molecular pathways than for academic biomedical research in general?

Research on EGF and NF-kB exhibits high levels of commercial activity, including patenting, while CTLA4 research is much closer to the norm for biomedical research. This is probably partly attributable to the more recent discovery of the functions of CTLA4 but not entirely; CTLA4, for whatever reasons, is not yet a target of much commercial interest.

For all three fields, respondents choose their project primarily on the basis of scientific importance, interest, feasibility, and funding. However, EGF investigators are more likely to cite personal income (11 percent versus 2 percent for the random academic sample) and the opportunity to start a new firm (7 percent versus 1 percent) as additional reasons to choose projects. Those working on NF-kB were above average in citing unreasonable terms for research inputs as a rea-

[22]Stephen Hansen, AAAS, presentation to the committee, February 11, 2005.

son to avoid pursuing a project (19 percent versus 10 percent for the norm). In none of the fields do "too many patents" appear to deter research.

The adverse effects of patents nevertheless occur more frequently for those who work on the pathways than for the random sample of academic biomedical researchers. Investigators working on the three pathways were two to three times more likely to indicate a need for access to a third-party patent than researchers in the random sample and were more likely to report adverse consequences. In CTLA4 research, there were no delays or modified projects, but one person abandoned a project. In EGF research, two researchers abandoned projects, three experienced delays, and one changed a research protocol. There were three reported NF-kB cases of delay and three of project redirection. Still the number of adverse incidents is small, representing less than 15 percent of the sample; and the number who had to abandon some project represents just 3 percent of those working on these pathways.

What are the effects of intellectual property on biomedical research and development in industry?

Presuming that patenting and commercialization are strong incentives to industrial research and development, especially in the biomedical arena, and that investigators would not report neglecting third-party intellectual property, the survey did not explicitly ask industry respondents about their reactions to upstream biomedical patents. However, a small number of industry respondents (17) answered related questions in which 60 percent said that they regularly check for third-party intellectual property and 35 percent acknowledged needing access to a third-party patent. Two out of the 17 said they had aborted a project for lack of such property, and 4 reported other adverse effects. It is unclear how many of these incidents were the result of being in direct market competition with the patent owner, but for this or other reasons it appears that the incidence of intellectual property-related problems is somewhat greater in industry than in academe.

Patents, Publications, and Citations

Fiona Murray and Scott Stern (2004) and Bhaven Sampat (forthcoming) have taken an entirely different approach in studying the effects of patents on scientific research and the anti-commons hypothesis regarding biomedical research in particular. Using slightly different methodologies, they examined what happens to the citations to a scientific article before and after a patent is issued on its subject matter. They found that articles associated with patents are more highly cited than articles not associated with patents, but that the citations are about 9 to 16 percent fewer than expected after the patent is awarded, suggesting some avoidance of the research direction and possibly some modest decline in "knowledge accumulation."

The finding is intriguing, especially in light of its corroboration by investigators using two different approaches. Nevertheless, for a host of methodological reasons, it should be interpreted with caution. Both papers' authors refrain from causal inferences or speculation about what lies behind their observations. Do investigators in fact know that a patent has issued? At least for academic researchers, this seems unlikely in view of the survey evidence that they neither search for patents nor respond to notices to pay attention to potential infringement. If they become aware of patents, do they cease working in an area or continue working but cite other research? In industry, where there is little premium on publication, the legal department often reviews external publications and may withhold them to avoid provoking patentees. In either case the effect, if real, may be more on publication and citation behavior than on research conduct. The effect, if real, ultimately may be more on citation behavior than on research conduct.

SHARING RESEARCH MATERIALS

In the meantime, the Walsh et al. survey turned up evidence of a more immediate and potentially remediable burden on research, private as well as public, stemming from difficulties in accessing proprietary research materials, patented or unpatented. Conflicts arising from scientific as well as commercial competition have to be addressed in addition to simply the burden and cost of providing such materials. Concern over the flow of research materials, which may be critical inputs for the success of a research project, is not new. Nor has it gone unaddressed; the NIH research tool guidelines address the process of materials exchanges, and NIH has developed a model Material Transfer Agreement (MTA).

The survey found that impediments to the exchange of biomedical research materials remain prevalent and may be increasing. For the period 1997 to 1999, Campbell and colleagues (2002) reported on the basis of a previous survey that academic genomics researchers denied 10 percent of material transfer requests. In the Walsh et al. study, the comparable number for 2003-2004 is 18 percent (95 percent confidence interval: +/- 3.7 percent). Other pertinent findings were as follows:

- Requests for material transfers between and within the industrial and academic sectors are widespread, although not of high frequency. About 60 percent of industry respondents and 75 percent of academic respondents initiated at least one request in the last two years. Approximately 40 percent of industry respondents and 69 percent of academic scientists had received such a request in the same period. Rates of initiation and receipt of requests are about the same for those doing drug discovery and those doing basic research.
- Between 7 percent (suppliers' estimate) and 18 percent (consumers' estimate) of university to university requests are denied. Typically, approximately half of respondents have had at least one request denied over a two-year period.

TABLE 4-4 Sharing of Research Materials, by Consumer Sector and Supplier Sector

Sectors		Average Percent Non-Compliance	
Consumer	Supplier	Consumer Estimate (%)	Supplier Estimate (%)
University	University	18	7
University	Industry	32	27
Industry	University	25	38
Industry	Industry	22	26

SOURCE: Walsh et al., 2005.

Rates of refusal or noncompliance are highest for university to industry, followed by industry to university and industry to industry requests (Table 4-4)

• The consequences of being denied a tangible research input can be more severe than the inability to license another's intellectual property, because in the latter case work may proceed, albeit at some liability risk. The survey asked about four possible adverse impacts—abandonment, delay, change in research approach, or the need to develop the research input in the requester's own laboratory. The results are shown in Table 4-5.

What stands out is the higher incidence of adverse effects for drug discovery and pathway researchers, especially those working on NF-kB.

• Fewer than half of material requests entail an MTA, and the presence or absence of a formal agreement does not appear to be central to whether ultimately the materials are shared. But the process of negotiating an MTA frequently entails costs in terms of restricted freedom of action, delays in proceeding, and financial costs to institutions. Reach-through claims are common as are publication restrictions, more so than royalty payments. Negotiations over MTA terms frequently occasion delays (with 11 percent of the requests leading to negotiations taking more than one month to conclude), especially when the suppliers are industrial firms. Industry suppliers also are much more likely to require MTA than academic suppliers.

• Although agreements for transfer of patented technologies are more likely to contain restrictive terms and have protracted negotiation histories than are agreements involving unpatented technologies, one cannot infer that patenting *per se* was the cause of the difficulties. Both patenting and complex drawn out negotiations derive from the commercial potential of the technology and the desire of the supplier and conceivably the consumer to capture a greater share of the rents from that potential.

• For academics, the most common reason given for denying or ignoring a request was simply the effort involved and the need to protect publication. For

TABLE 4-5 Average Number of Adverse Effects from Not Receiving Research Inputs, Academic Respondents by Research Goal and for Pathways and Industry Respondents

		Research Goal			Pathways			
	Random Sample	Drug Discovery	Basic Research	Other	CTLA4	EGF	NF-kB	Industry
			Academic Supplier					
Delay >1 month	0.68	0.98	0.69	0.33	0.83	1.2	2.85	0.78
Change Research Approach	0.56	0.89	0.54	0.3	0.45	0.7	2.24	0.68
Abandon	0.22	0.07	0.24	0.21	0.27	0.2	0.62	0.39
Make In-house	0.67	0.88	0.65	0.59	0.93	1.2	2.29	1.01
			Industry Supplier					
Delay >1 month	0.4	0.75	0.39	0.18	1.02	1.1	0.87	0.35
Change Research Approach	0.46	0.66	0.42	0.56	0.68	0.7	1.66	0.49
Abandon	0.27	0.08	0.3	0.26	0.58	0.9	0.28	0.32
Make In-house	0.31	0.44	0.28	0.47	0.69	0.8	0.71	0.33
Respondents	242	24	195	23	21	24	26	62

SOURCE: Walsh et al., 2005.

industry respondents, the key reported reasons were the need to protect commercial value and the consumer's unwillingness to accept restrictive terms.

Gene-Based Diagnostic Test Patents

An area where patents seem to be having an inhibitory effect on research and related clinical practice involves gene-based diagnostic tests (see Chapter 2 for a discussion of breast cancer diagnostics). This was not a focus of the committee's survey, in part because Mildred Cho and colleagues (2003) have conducted telephone surveys of U.S. clinical laboratory directors who were members of the Association for Molecular Pathology. The first concern is that a patent owner's refusal to make a single patented gene available for licensing on reasonable terms will inhibit follow-on research on the incidence of mutations in the gene as well as limit patient access to testing at a reasonable cost and the possibility of obtaining a second opinion on the result. Exclusive licenses also limit the opportunity for the development of improvements in the test and verification of the result. An anti-commons effect also can be anticipated in the future as clinicians begin to develop tests for multigenic traits.

Cho and colleagues' sampling frame of 211 laboratory directors combined listings in the most recent Association of Molecular Pathology directory and on the *genetests.org* website maintained by the University of Washington with federal funding. The result was a sample of corporate, university, private hospital, federal government, and other nonprofit laboratories. They analyzed the responses of 122 individuals, a large majority of whom had licenses to perform genetic tests for a wide variety of conditions, including hereditary breast and ovarian cancer (BRCA1/2), Canavan Disease, Hereditary Hemochromatosis, and Fragile X syndrome, among others.

The results suggest that holders of gene-based diagnostic patents are active in asserting their intellectual property rights. Sixty-five percent of respondents reported having been contacted by a patent or license holder regarding their potential infringement in performing a test. Twenty laboratories had received notification for 1 test; 51 had received notifications for up to 3 tests, and 26 laboratories for 4 or more tests. These enforcement efforts focused on 12 tests that, as a result, 1 or more laboratories had ceased to perform. In all, 30 laboratories responded that they had ceased administering at least 1 test. This number included almost all of the corporate laboratories and about one-quarter of university laboratories. Asked to evaluate their experience, respondents indicated that patents had had a negative effect on all aspects of clinical testing and reported a decline in information sharing between laboratories. Inclination to undertake test development, too, was adversely affected, according to respondents. Thus, patents do appear to be blocking the clinical use of tests insofar as such clinical use is closely related to follow-on research. Because clinical research often is more efficiently done with an entire battery of tests, both blocking and an anti-commons might be in effect.

CONCLUSION

After reviewing the existing research literature and conducting research on (1) issued patents and published patent applications in a subset of biotechnology categories; (2) a small set of university licensing practices of selected categories of patents; and (3) biomedical research scientists' experiences with intellectual property and its effects on research, the committee identified four areas of concern.

First, the apparent lack of substantial evidence for a patent thicket or a patent-blocking problem is associated with a general lack of awareness or concern among investigators about existing intellectual property. This situation could change dramatically as institutions increasingly realize that they enjoy no legal protection and concerns are raised about possible patent infringement liability; this may lead them to take more steps to raise awareness and regulate their behavior.

Second, although the survey did not reveal significant differences in experience between investigators working independently and those working in multimember teams, the growing complexity of biomedical research may make intellectual property more problematic as work on a single gene or gene sequence gives way to research entailing far more extensive inputs, more and more of them patented.

Third, the licensing of some gene-based diagnostic tests does appear to be having an inhibiting effect on research and related clinical practice.

Finally, impediments to the exchange of research materials among laboratories exist, although these impediments appear to be largely independent of intellectual property. Instead, they are associated with scientific competition and the lack of rewards for the time and effort entailed in meeting requests for research inputs and academic respondents' commercial interests.

5

Conclusions and Recommendations

The previous chapters have described how the nature of molecular biology and the behavior norms of the scientific community have changed in the wake of the Human Genome Project (HGP) and the birth of proteomics. A complement to the traditional hypothesis-driven study of single genes or proteins is the option of simultaneously studying many genes or proteins. This sea change in the field has occurred while both university and private sector scientists have been aggressively protecting intellectual property of discoveries that are well upstream of practical applications. Thus, the potential exists where discoveries in genomics and proteomics that will benefit the public health and well-being could be thwarted by complex intellectual property problems. In Chapter 4, the committee's findings from its own research, as well as that of others, on how intellectual property practices and enforcement are affecting genomics and proteomics research are presented. In this chapter, the committee draws conclusions and makes recommendations in three overarching areas that aim to ensure that the public investment in genomics and proteomics results in optimal public benefit:

1. improving and facilitating best practices and norms in the conduct of genomics and proteomic research;
2. adapting the patent system to the rapidly changing fields of genomics and proteomics; and
3. facilitating research access to patented inventions through licensing and shielding from liability for infringement.

OVERALL CONCLUSIONS

The committee found that the number of projects abandoned or delayed as a result of technology access difficulties is reported to be small, as is the number of occasions in which investigators revise their protocols to avoid intellectual property complications or pay high costs to obtain access to intellectual property. Thus, for the time being, it appears that access to patents or information inputs into biomedical research rarely imposes a significant burden for academic biomedical researchers. However, for a number of reasons, the committee concluded that the patent landscape, which already is burgeoning in areas such as gene expression and protein-protein interactions, could become considerably more complex and burdensome over time.

There are reasons to be concerned about the future. First, the lack of substantial evidence for a patent thicket or a patent blocking problem clearly is linked to a general lack of awareness or concern among academic investigators about existing intellectual property. That could change dramatically and possibly even abruptly in two circumstances. Institutions, aware that they enjoy no protection from legal liability, may become more concerned about their potential patent infringement liability and take more active steps to raise researchers' awareness or even to try to regulate their behavior. The latter could be both burdensome on research *and* largely ineffective because of researchers' autonomy and their ignorance or at best uncertainty about what intellectual property applies in what circumstances. Alternatively, patent holders, equally aware that universities are not shielded from liability by a research exception, could take more active steps to assert their patents against them. This may not lead to more patent suits against universities—indeed, established companies are usually reluctant to pursue litigation against research universities—but it could involve more demands for licensing fees, grant-back rights, and other terms that are burdensome to research. Certainly, some holders of gene-based diagnostic patents are currently active in asserting their intellectual property rights. Even if neither of these scenarios materializes, researchers and institutions that unknowingly and with impunity infringe on others' intellectual property could later encounter difficulties in commercializing their inventions.

Finally, as scientists increasingly use the high-throughput tools of genomics and proteomics to study the properties of many genes or proteins simultaneously, the burden on the investigator to obtain rights to the intellectual property covering these genes or proteins could become insupportable, depending on how broad the scope of claims is and how patent holders respond to potential infringers. The large number of issued and pending patents relating to gene-expression profiling and protein-protein interactions contributes to this concern.

More immediately, the survey data revealed substantial evidence of another, potentially remediable burden on private as well as public research stemming from difficulties in accessing proprietary research materials, whether patented or unpatented. The committee found that impediments to the exchange of biomedical research materials remain prevalent and may be increasing.

Several steps can be taken to prevent an increasingly problematic environment for research in genomics and proteomics as more knowledge is created, more patent applications are filed, and more restrictions are placed on the availability of and access to information and resources.

BEST PRACTICES AND NORMS FOR THE SCIENTIFIC COMMUNITY AND FEDERAL RESEARCH SPONSORS

Many of the potential problems looming in the realm of genomics, proteomics, and intellectual property can be avoided if scientists and their institutions, whether public or private, follow the best practices already articulated by the National Institutes of Health (NIH), the National Research Council (NRC), and others. U.S. science has flourished because of its general openness and the sharing of data and research resources. This is not to suggest that legitimate proprietary interests in science do not exist, but rather is intended to highlight the argument that whenever possible, sharing is in the best interest of all science, both basic and applied. Several measures can be taken to facilitate the free exchange of materials and data.

Foster Free Exchange of Data, Information, and Materials

From the inception of the HGP, public and commercial funders of these activities have emphasized that, in order to reap the maximum benefit to the public health, data should be freely available in the public domain. In addition, the NRC has repeatedly emphasized the need for sharing data. The council's 2003 report *Sharing Publication-Related Data and Materials* endorsed the uniform principle for sharing integral data and materials expeditiously:

> Community standards for sharing publication-related data and materials should flow from the general principle that the publication of scientific information is intended to move science forward. More specifically, the act of publishing is a *quid pro quo* in which authors receive credit and acknowledgement in exchange for disclosure of their scientific findings. An author's obligation is not only to release data and materials to enable others to verify or replicate published findings but also to provide them in a form on which other scientists can build with further research. All members of the scientific community—whether working in academia, government, or a commercial enterprise—have equal responsibility for upholding community standards as participants in the publication system, and all should be equally able to derive benefits from it (NRC, 2003, p. 4).

Nucleic acid sequences provide the fundamental starting point for describing and understanding the structure, function, and development of genetically diverse organisms. For almost 20 years, GenBank, the European Molecular Biology Laboratory, and the DNA Data Bank of Japan have collaborated to create nucleic acid sequence data banks. These data banks are invaluable to researchers but they face insufficiencies and gaps as fewer data deposits are made because of proprietary interests.

The genomics and proteomics communities, in general, have honored these calls for data sharing, especially in the large-scale projects such as the HGP itself, the Expressed Sequence Tag (EST) project, and the SNP Consortium. Some practices, however, do not conform to these norms. Private industry consistently retains some portion of its protein structure information in proprietary databases, effectively withholding from the scientific community a large and important dataset that could facilitate basic and applied research in structural biology. However, once structures are no longer commercially important, their availability in the public domain would be very useful for academic research. In addition, defensive patenting of three-dimensional structures of drug targets has the potential to interfere with drug discovery. Structural biology data in published patent applications and issued patents are presented in such a form that they are not readily incorporated into the Protein Data Bank (PDB) for the benefit of the larger scientific community. Furthermore, academic scientists are sometimes driven by competitive pressures to withhold both information and materials.

Eventually, large-scale structural genomics efforts will dominate the production of new structures. Full disclosure of structures without patenting could serve to preempt much of the defensive patenting currently sought by industry and substantially improve the environment for all of science. The committee commends NIH for its effective use of provisions in Requests for Proposals for projects involving the development of resources for the public domain that require that grant applicants include in their proposals an explanation of their plans for the sharing and dissemination of research results. Although NIH does not currently collect and analyze data on grantee behavior, it has the ability and the authority to elicit good behavior among grantees and contractors and should exercise that authority wherever possible.

Recommendation 1:

NIH should continue to encourage the free exchange of materials and data. NIH should monitor the actions of grantees and contractors with regard to data and material sharing and, if necessary, require grantees and contractors to comply with their approved intellectual property and data sharing plans.

However, it should be noted that investigators have the right and even the obligation to retain materials and data until they are confident of their validity and have reported their results in publication. The quality of science and the value of the public data must be upheld even while meeting the goal of sharing materials and data.

Recommendation 2:
The committee supports NIH in its efforts to adapt and extend the "Bermuda Rules" to structural biology data generated by NIH-funded centers for large-scale structural genomics efforts, and thereby making data promptly and freely available in a database via the PDB.

Although in principle the coordinate data that are in patent applications could be put into the PDB, both the content and format of these patent applications are not suitable for incorporation into the repository. The PDB has established standard formats for electronically archiving the coordinate, experimental, and meta data. Recently USPTO proposed that these data be sent in electronic form as part of relevant patent applications. The Worldwide PDB, an international organization responsible for all PDB data, endorsed this proposal and further stipulated that the standard formats be required. This would ensure that the data would be efficiently and properly archived and be made freely available.

Recommendation 3:
The PDB should work with USPTO, the European Patent Office (EPO), and the Japanese Patent Office (JPO) to establish mechanisms for the efficient transfer of structural biology data in published patent applications and issued patents to the PDB for the benefit of the larger scientific community. To the extent feasible within commercial constraints, all researchers, including those in the private sector, should be encouraged to submit their sequence data to GenBank, the DNA Databank of Japan, or the European Molecular Biology Laboratory and to submit their protein structure data to the PDB.

Foster Responsible Patenting and Licensing Strategies

In 1999, NIH issued *Principles and Guidelines for Recipients of NIH Research Grants and Contracts on Obtaining and Disseminating Biomedical Research Resources* (64 FR 72090).[1] These aspirational principles were issued by NIH to provide guidance and direction to NIH-funded institutions in order to balance the need to protect intellectual property rights with the need to broadly disseminate new discoveries. They recognize that licensing policies and practices are extremely important determinants of the effects of patents on upstream technologies on the conduct of follow-on research. The principles apply to all NIH-funded entities and address biomedical materials, which are broadly defined to include cell lines, monoclonal antibodies, reagents, animal models, combinatorial

[1] A copy of the complete principles can be obtained at the NIH Web site at *http://www.nih.gov/od/ott/RTguide_final.htm.*

chemistry libraries, clones and cloning tools, databases, and software (under some circumstances).[2]

The principles were developed in response to complaints from researchers that restrictive terms in material transfer agreements (MTAs) were impeding the sharing of research resources. These restrictions came both from industry sponsors and from research institutions. In the *Principles and Guidelines*, NIH urges recipient institutions to adopt policies and procedures to encourage the exchange of research tools by minimizing administrative impediments, ensuring timely disclosure of research findings, ensuring appropriate implementation of the Bayh-Dole Act, and ensuring the dissemination of research resources developed with NIH funds.

Consistent with its ongoing interest in facilitating broad access to government-sponsored research results, NIH in 2004 issued *Best Practices for the Licensing of Genomic Inventions*. This document aims to maximize the public benefit whenever technologies owned or funded by the Public Heath Service are transferred to the commercial sector. In this document, NIH recommends that "whenever possible, non-exclusive licensing should be pursued as a best practice. A non-exclusive licensing approach favors and facilitates making broad enabling technologies and research uses of inventions widely available and accessible to the scientific community." The document goes on to say that "exclusive licenses should be appropriately tailored to ensure expeditious development of as many aspects of the technology as possible." The policy distinguishes between diagnostic and therapeutic applications and cautions against exclusive licensing practices in some areas. For example, the document states that "patent claims to gene sequences could be licensed exclusively in a limited field of use drawn to development of antisense molecules in therapeutic protocols. Independent of such exclusive consideration, the same intellectual property rights could be licensed non-exclusively for diagnostic testing or as a research probe to study gene expression under varying physiological conditions."[3]

The committee endorses these NIH policies, in particular the principles that patent recipients should analyze whether further research, development, and private investment are needed to realize the usefulness of their research results and that proprietary or exclusive means of dissemination should be pursued only when there is a compelling need. Also, whenever possible, licenses should be limited to relatively narrow and specific commercial application rather than as blanket exclusive licenses for uses that cannot be anticipated at the moment.

---

[2]The guidelines were issued following recommendations made to the NIH Advisory Committee to the Director by a special subcommittee chaired by Rebecca Eisenberg.

[3]On April 11, 2005, NIH published the final notice, after receipt of public comments, at *http://ott.od.nih.gov/lic_gen_inv_FR.html*.

Recommendation 4:
The committee endorses NIH's *Principles and Guidelines for Recipients of NIH Research Grants and Contracts on Obtaining and Disseminating Biomedical Research Resources* and *Best Practices for the Licensing of Genomic Inventions*. Through its *Guide for Grants and Contracts*, NIH should require that recipients of all research grant and career development award mechanisms, cooperative agreements, contracts, institutional and Individual National Research Service Awards, as well as NIH intramural research studies, adhere to and comply with these guidance documents. Other funding organizations (such as other federal agencies, nonprofit and for-profit sponsors) should adopt similar guidelines.

These principles can and should be followed by other funding agencies. In addition, they should be followed consistently for gene patents and licenses, and they should be applied to proteomics research to discourage inappropriate patenting and licensing practices. For example, the committee believes that it would be consistent with the NIH guidelines to discourage grantees and contractors from patenting three-dimensional macromolecular structures. For the sake of clarity, the committee does not believe that NIH grantees and contractors should be discouraged from patenting biological macromolecules that have been shown to have clear therapeutic value in their own right. The committee recognizes the value of patents when follow-on private investment adds social value by bringing products and services to market, and while this is to be commended, licensing should be done in ways that permit continued research and avoid logjams, undue royalty stacking, and anti-commons problems.

Because NIH issued these policies as guidance documents, grantees and contractors are not required to comply with them. Nor are researchers and research institutions not funded by NIH under any obligation to comply. The committee believes that NIH should continue to encourage adherence to these guidelines and best practices by the extramural community. However, in circumstances in which grantees are found to be ignoring the guidelines and thereby inhibiting innovation, the committee believes that NIH should use its authority to make adherence to the guidelines a condition of a future grant or contract award. By placing the responsibility with the applicant, NIH can state a position relative to its overall goal, but not generate endless pages of detailed policies and procedures. This is an evolving area where flexibility is important. If the goal is normative behavior, some process must be in place to make institutions and investigators examine their own behavior and articulate how they will behave in the broad context of agreed-upon goals. If those positions are widely shared, as in the grant application process, they will help to develop consensus about acceptable or desirable behavior. If there is flexibility in how institutions can approach these issues, then the entire field will reap the benefit of creative approaches.

In addition, NIH and the broader research community should encourage, wherever possible, voluntary compliance with the intent of these policy documents. There are many precedents for voluntary compliance with such standards by industry, dating back to the voluntary submission of research protocols involving recombinant DNA, and more recently, gene transfer studies, to NIH's Recombinant DNA Advisory Committee for review.

Furthermore, the committee's research found that most institutions report that they reserve rights for their own investigators to use a patented technology even though it is licensed exclusively to a commercial entity. An increasingly common university practice in recent years is to reserve such rights for investigators at other nonprofit institutions, but this often is subject to the patent holder's case-by-case approval. The committee commends and endorses this practice, which could be applied to other organizations, as appropriate.

Recommendation 5:
Universities should adopt the emerging practice of retaining in their license agreements the authority to disseminate their research materials to other research institutions and to permit those institutions to use patented technology in their nonprofit activities.

In addition, to support the dissemination of biological research materials to the scientific research community, institutions use Material Transfer Agreements (MTAs) in handling the exchange of research materials with the research community. MTAs are intended to protect the institution's ownership interest in the research material and contain provisions regarding the distribution and use of the research material. However, in the committee's opinion, the use and complexity of these agreements have become burdensome and overly restrictive. Institutions should promote the exchange of material and data while protecting legitimate intellectual property interests.

Recommendation 6:
In cases in which agreements are needed for the exchange of research materials and/or data among nonprofit institutions, researchers and their institutions should recognize restrictions and aim to simplify and standardize the exchange process. Agreements such as the Simple Letter Agreement for the Transfer of Materials or the Uniform Biological Material Transfer Agreement (UBMTA) can facilitate streamlined exchanges. In addition, NIH should adapt the UBMTA to create a similar standardized agreement for the exchange of data. Industry is encouraged to adopt similar exchange practices.

ADAPTING THE PATENT SYSTEM TO THE DEVELOPING FIELDS OF GENOMICS AND PROTEOMICS

To obtain a patent an applicant must claim an invention that falls within patent-eligible subject matter. The invention must be new, useful, and nonobvious in light of the prior art. The patent application must satisfy certain disclosure requirements, including a written description of the invention, an enabling disclosure that allows a person of ordinary skill in the field to make and use the invention without undue experimentation, and disclosure of the best mode contemplated by the inventor of carrying out the invention. The exclusion of abstract ideas from patent protection traditionally has been more important for information technology than for biotechnology, but some genomics and proteomics research has the potential to confuse or even to blur the boundaries between abstract ideas and applications.

The fields of genomics and proteomics are dependent on rapidly changing technology and complex theory. Understanding biological processes through the association of genetic variation with individual phenotypic differences and through structural analyses will involve a variety of methods (global medical sequencing and population genetics in the first and x-ray crystallography and nuclear magnetic resonance [NMR] spectroscopy in the second). These methods will raise many new challenges for USPTO and the courts.

The challenge of these types of innovations clearly was illustrated in the 1990s when the scientific community was in intense discussions with USPTO about the value of ESTs. It will be increasingly important for patent examiners to be current with scientific and clinical developments in the field.

Recommendation 7:
USPTO should create a regular, formal mechanism, such as a chartered advisory committee or a regularly scheduled forum, comprising leading scientists in relevant emerging fields, to inform examiners about new developments and research directions in their field. NIH and other relevant federal research agencies should assist USPTO in identifying experts to participate in these consultations.

USPTO is to be commended for the development of its Customer Partnership Program for biotechnological patent applications. The committee urges USPTO to expand the use of input from the scientific community to improve the understanding of the office and its examiners of complex and rapidly evolving technologies, such as genomics and proteomics, with both human health and agricultural applications. The proposed committee should follow the Federal Advisory Committee Act requirements for open meetings and advance notification of meetings.

Nonobviousness

As described in Chapter 3, the *In re Bell* decision is illustrative of the application to genomics of the requirements for nonobviousness. In that case, USPTO argued that a defined gene sequence was obvious from prior art, including the sequence of the encoded protein and a general method of cloning. The inventor argued that the prior art relied upon by USPTO did not suggest all of the modifications to the cited cloning technique that would make it operative and that USPTO had, without supporting evidence, deemed such modifications within the ordinary skill of the field.

In *Bell* and then *In re Deuel* the court held that—as of the time the invention was made—a gene is just another type of chemical compound, and the issue for nonobviousness is the structure (that is, sequence) of the gene. Unless the sequence is predictable from the prior art, the gene is nonobvious. In these two cases, the court refused to see that there is a known relationship between a gene and the protein it encodes.

The National Academies' 2004 report, *A Patent System for the 21st Century,* observed that advances in proteomics have shown that the relationship between DNA sequence and protein *sequence* is predictable, but the relationship to the *structure* of the protein is not. The report noted that newly disclosed protein structures might satisfy the nonobviousness standard more easily than newly disclosed DNA molecules, given that the fine details of the three-dimensional structures cannot be deduced accurately from either the protein or DNA sequence. On the other hand, as more proteomic information becomes publicly available through large-scale projects, the ability to predict the structure based on the amino acid sequence of a protein and the ease with which the structure is obtained will dramatically improve. Nonobviousness determinations require that one look to the prior art and assess whether a person of ordinary skills could replicate the invention, whether such a person would be motivated to do so, and whether he or she would have a reasonable chance of success.

The previous National Academies' committee recommended that the Federal Circuit abandon the rule announced in *Bell* and *Deuel* that, essentially, prevents the consideration of the technical difficulty faced in obtaining pre-existing genetic sequences. The National Academies sought an approach similar to that of other industrialized countries when examining the obviousness of gene-sequence-related inventions: Each case should be analyzed at least in part by looking at the technical difficulty a skilled artisan would have faced at the time the invention was discovered.

Recommendation 8:
In determining nonobviousness in the context of genomic and proteomic inventions, USPTO and the courts should avoid rules of nonobviousness that base allowances on the absence of structurally similar molecules and instead should evaluate obviousness by considering whether the prior art indicates that a scientist of ordinary skill would have been motivated to make the invention with a reasonable expectation of success *at the time the invention was made.*

NIH should partner with other organizations (e.g., the Federal Judicial Center) to develop venues for educating judges about advances and new developments in the areas of genomics and proteomics.

Utility Standard

The Supreme Court articulated a strict utility standard in its 1966 decision in *Brenner* v. *Manson*, requiring that a patent applicant show that the invention has "specific benefit in currently available form." The court justified this strict approach by noting that "a patent is not a hunting license. It is not a reward for the search, but compensation for its successful conclusion." But the standard has not been applied in a consistent fashion. Some believe more recent decisions of the Federal Circuit have been less strict about the utility requirement, particularly as applied to biopharmaceutical inventions.

The 2002 report on a trilateral comparative study by the EPO, the JPO, and USPTO (2002 trilateral report) considers the patentability of claims related to the three-dimensional structure of proteins under the laws administered by each of those offices. Each of the three concluded that hypothetical claims to computer models of proteins generated with atomic coordinates, data arrays comprising the atomic coordinates of proteins, computer-readable storage medium encoded with the atomic coordinates, and databases encoded with candidate compounds that had been electronically screened against the atomic coordinates of proteins were not patent eligible. The analysis by USPTO emphasized that each of these hypothetical claims was "nonfunctional descriptive material" and therefore "an abstract idea."

Understanding how genetic variation leads to individual variation in humans is one of the great scientific challenges of the twenty-first century. The path forward will inevitably involve an increasingly broad survey of genetic variation across the genome and establishing the causal relationship of certain regions and ultimately genes with particular traits. Indeed, technology already is in development that would allow complete cataloging of an individual's genetic code at affordable costs. As these technologies are implemented, diagnostics will move from a focus on single genes to a search of all genes.

If it is determined to be essential to allowing research to proceed and medical practice to advance in the coming years, those who are discovering associations between sequence variants and traits should eschew patents. Failing that, the best practices established by NIH and the broader scientific community should be followed. USPTO should require high standards for utility as mandated by existing Supreme Court precedent.

Although the views of USPTO and its foreign counterparts are of enormous practical importance in determining what receives a patent, neither the USPTO guidelines nor the 2002 trilateral report has the status of binding legal authority. As discussed in Chapter 3, the utility standard has proven difficult to administer in a consistent fashion. The committee believes this problem should be addressed.

The committee endorses the USPTO utility and written description guidelines and commends the office for adopting them. The committee also commends the process of input from the scientific community that led to their adoption and modification. Ongoing dialogues of this sort, and as recommended above, should form the basis for continually adapting the guidelines as the underlying science moves forward. However, the scientific community also must bear some responsibility for interpreting the guidelines.

Recommendation 9:
Principal Investigators and their institutions contemplating intellectual property protection should be familiar with the USPTO utility guidelines and should avoid seeking patents on hypothetical proteins, random single nucleotide polymorphisms and haplotypes, and proteins that have only research, as opposed to therapeutic, diagnostic, or preventive, functions.

A move toward a higher standard by the scientific community, USPTO, and the courts would be consistent with the 2001 USPTO guidelines initially adopted to limit patenting of ESTs. Those guidelines recently have been upheld by the Court of Appeals for the Federal Circuit (*In re Fisher*). The committee believes that such guidelines have had a beneficial effect and USPTO should ensure that they are applied to proteomic inventions.

FACILITATE RESEARCH ACCESS TO PATENTED INVENTIONS THROUGH LICENSING AND SHIELDING FROM LIABILITY FOR INFRINGEMENT

Experimental Use Exemption

Academic scientists commonly assume that their research is shielded by law from intellectual property infringement liability (NRC, 1997). However, in *Madey* v. *Duke University*, the Federal Circuit rejected the experimental use defense in the context of academic research, declaring the noncommercial character of the research to be irrelevant to its analysis of the case. The court found that research that is part of the "legitimate business" of the university is not protected "regardless of commercial implications" or lack thereof.[4] The implications of this decision are not yet clear, although it would appear that researchers and their institutions will have to pay much closer attention to the intellectual property issues involved in their current and future work especially when that work is driven by commercial considerations. Given the nature of much university research—that

[4]*Madey* v. *Duke University*, 307 F.3d 1351 (Fed.Cir. 2002).

is, investigator initiated, highly decentralized, and uncoordinated—the implementation of an administrative structure that would deal prospectively with intellectual property issues in a manner similar to due diligence precautions in the private sector could impose burdensome administrative costs and strongly influence choices of academic research directions. At the same time, it is doubtful that such an apparatus could be effective in a university context. The ongoing "research exception" litigation is indicative that many aspects of the law governing patent rights to research tools are not settled.

The committee believes that there should be a statutory exemption from infringement for experimentation *on* a patented invention.

Recommendation 10:
Congress should consider exempting research "on" inventions from patent infringement liability. The exemption should state that making or using a patented invention should not be considered infringement if done to discern or to discover:

a. the validity of the patent and scope of afforded protection;
b. the features, properties, or inherent characteristics or advantages of the invention;
c. novel methods of making or using the patented invention; or
d. novel alternatives, improvements, or substitutes.

Further making or using the invention in activities incidental to preparation for commercialization of noninfringing alternatives also should be considered noninfringing. Nevertheless, a statutory research exemption should be limited to these circumstances and not be unbounded. In particular, it should not extend to unauthorized use of research tools for their intended purpose, in other words, to research "with" patented inventions. Accordingly, our recommendation would not address the circumstances of the *Madey* case, which clearly entailed research "with" the patented laser; but it would shield some types of biomedical research involving patented subject matter.

Patent Pooling

A patent pool is an agreement between two or more patent owners to license one or more of their patents to one another or third parties.[5] A 2000 white paper issued by USPTO promoted their use, stating:

> The use of patent pools in the biotechnology field could serve the interests of the public and private industry, a win-win situation. The public would be served by having ready access with streamlined licensing conditions to a greater amount of

[5]See Klein, *supra at http://www.usdoj.gov/atr/public/speeches/1123.html.*

> proprietary subject matter. Patent holders would be served by greater access to licenses of proprietary subject matter of other patent holders, the generation of affordable pre-packaged patent "stacks" that could be easily licensed, and an additional revenue source for inventions that might not otherwise be developed. The end result is that patent pools, especially in the biotechnology area, can provide for greater innovation, parallel research and development, removal of patent bottlenecks, and faster product development (USPTO, 2000, p. 11).

The committee agrees that patent pooling is an approach that might address some issues of access to patented upstream technology and its possible applications to biomedical research and development and that it should be studied.

Recommendation 11:
NIH should undertake a study of potential university, government, and industry arrangements for the pooling and cross-licensing of genomic and proteomic patents, as well as research tools.

Such proposed sharing arrangements are being pursued in agricultural biotechnology by the Public Intellectual Property Resource for Agriculture and the Biological Innovation for Open Society initiative in different ways. One issue that may be important in the lucrative health field is the willingness of academic scientists to have their inventions pooled if that would reduce or threaten their receipt of the share of royalties as typically are provided by universities.

Ensuring the Public's Health

Although the committee was unable to find any evidence of systematic failure of the licensing system, a few cases of restrictive or refusals to license practices by some companies have generated controversy and disapproval because of the potential adverse effects on public health. Through the Agreement on Trade-Related Aspects of Intellectual Property Rights (TRIPS Agreement), some other countries, such as Belgium and Canada, retain the right to issue compulsory licenses if there is a public health imperative. In the United States, courts have used their equitable powers to deny injunctive relief in cases where health and safety are in issue.[6]

Although this option is rarely used and difficult to implement, the threat that a court might decline to enforce a patent by enjoining its infringement may be enough to spur patent holders to license on reasonable terms (OECD, 2002). It

---

[6]See, e.g., *Roche Products, Inc. v. Bolar Pharmaceuticals Co.*, 733 F.2d 858, 866 (Fed. Cir. 1984); *Vitamin Technologists, Inc. v. Wisc. Alumni Res. Found.*, 146 F.2d 941, 956 (9th Cir. 1945); *City of Milwaukee v. Activated Sludge, Inc.*, 69 F.2d 577, 593 (7th Cir. 1934).

always should be a last resort, when all else fails, and when protection of the public health cannot be achieved by any other means.

Recommendation 12:
Courts should continue to decline to enjoin patent infringement in those extraordinary situations in which the restricted availability of genomic or proteomic inventions threatens the public health or sound medical practice. Recognition that there is no absolute right to injunctive relief is consistent with U.S. law and with the Agreement in Trade-Related Aspects of Intellectual Property Rights (the TRIPS Agreement).

Gene-Based Diagnostic Testing

Absent special circumstances, such as when the costs of development are high, the licensing of genomic and proteomic tools should be broad so that they ensure patient access and the opportunity to improve upon the method. The committee recognizes that diagnostic tests will sometimes involve such special circumstances and that there is a need to license more exclusively when the costs of test development or diffusion require the substantial investment of private capital. It is likely that with continued advancements in human genomics and the recognition of ever more statistical correlations between mutations in multiple genes and clinical phenotypes, opportunities for engaging in such restrictive practices will continue to multiply. Nevertheless, licenses on genomic- or proteomic-based diagnostic tests, when inventing around the test is not possible, should create reasonable access for patients, allow competitive perfection of the test, not interfere with noncommercial applications of the test in Institutional Review Board (IRB)-approved clinical research, and ensure compliance with regulatory requirements such as permitting quality verification. To ensure a reasonable return on investment, the license may require that the licensee first be given the opportunity to furnish the materials or services required.

The committee recognizes that exclusivity is commonly required to secure the large amounts of investment capital that are needed to establish testing capability on an industrial scale. On the other hand, the exclusive practice of any medical procedure or clinical diagnostic test is an important issue for the medical profession and raises important questions of public health and science policy. For example, the performance of a gene-based clinical test in an academic setting often generates rich databases of newly detected genetic variations that can be correlated with an array of clinical phenotypes. Such admixed medical practice and research provides important new information about the mutational repertory of specific disease-linked genes, as well as the phenotypic correlations that provide new insights into disease mechanisms and identify potential new targets for therapeutic intervention. In instances of the exclusive patenting or licensing of a test, such correlations will only occur if the data derived from the test are made freely available to the clinicians treating the patients. Thus, clinical research in the United States always has been intertwined with the practice of medicine by physician investigators in academic medical institutions, and historically, overages obtained from medical practice have been a significant source for investment and operating funds in clinical research.

Furthermore, the practice of gene-based diagnostic tests by academic laboratories on the large and heterogeneous patient populations of the academic medical center generates rich databases of newly detected genetic variations that can be correlated with an array of clinical phenotypes. Such admixed medical practice and research provides important new information about the mutational repertory of specific disease-linked genes, as well as the phenotypic correlations that provide new insights into disease mechanisms and identify potential new targets for therapeutic intervention. Such research is a hallmark of academic medical practice and historically has made enormous contributions to the advancement of medical knowledge and public health.

It also is the case that health professionals, the biopharmaceutical industries, and the public are anticipating eagerly a new era of "individualized medicine" and the application of pharmacogenomics to guide the drug development process and tailor therapeutic interventions to individuals and populations based on known genetic factors predictive of drug efficacy and safety. For industry to exploit this promising potentiality, the development and practice of precise, gene-based diagnostic tests to identify the candidate populations for both drug testing and marketing will be required. The development of new genetic tests will be linked intimately as never before to drug development, testing, and marketing.

Given the rapid development of gene-based diagnostic testing and its increasingly critical role in the practice of medicine, the committee identified a variety of concerns that it believes should be considered in licensing practices on genomic- or proteomic-based diagnostic tests, where inventing around the test may not be possible, including:

- access for patients;
- allowing competitive perfection of the tests;
- facilitating IRB-approved clinical research in academic medical centers regardless of funding sources;
- facilitating professional education and training;
- permitting independent validation of test results; and
- ensuring regulatory compliance.

Although the committee discussed all of the above concerns at length, it was especially concerned with independent validation of genomic- or proteomic-based test results. Certain members of the medical and academic community noted that, where patent owners may control access to genomic- or proteomic-based diagnostic tests, the patent owners may not allow others to use the patented technologies to validate the results of particular clinical tests. The committee agreed that this may present a problem and encourages patent owners to consider entering into licenses that will permit others to use the patented technologies for the purpose of independently confirming the results of a diagnostic test.

Recommendation 13:
Owners of patents that control access to genomic- or proteomic-based diagnostic tests should establish procedures that provide for independent verification of test results. Congress should consider whether it is in the interest of the public's health to create an exemption to patent infringement liability to deal with situations where patent owners decline to allow independent verification of their tests.

References

Abate, T. 1999. Biotech firms rushing to patent gene fragments. *The San Francisco Chronicle.* October 18. p. B14.

Alberts, B., and Sir A. Klug. 2000. The human genome itself must be freely available to all humankind. *Nature.* 404(6776):325.

Atkinson, R. C., R. N. Beachy, G. Conway, F. A. Cordova, M. A. Fox, K. A. Holbrook, D. F. Klessig, R. L. McCormick, P. M. McPherson, H. R. Rawlings III, R. Rapson, L. N. Vanderhoef, J. D. Wiley, and C. E. Young. 2003. Public sector collaboration for agricultural IP management. *Science.* 301(5630):174-175.

Avery, O. T., C. M. MacLeod, and M. McCarty. 1944. Studies on the chemical nature of the substance inducing transformation of Pneumococcal types: induction of transformation by a Deoxyribonucleic Acid fraction isolated from Pneumococus Type III. *J. Exp. Med.* 79(2):137-157.

Berg, J. M. 2004. Presentation to Committee on Intellectual Property in Genomic and Protein Research and Innovation at February 11-12 meeting.

Blumenthal, D., E. G. Campbell, M. S. Anderson, N. Causino, and K. S. Louis. 1997. Withholding research results in academic life science: evidence from a national survey of faculty. *JAMA.* 277(15):1224-1228.

Blumenthal, D., E. G. Campbell, N. Causino, and K. S. Louis 1996. Participation of life-science faculty in research relationships with industry. *N. Engl. J. Med.* 335(23):1734-1739.

Blumenthal, D., M. Gluck, K. S. Louis, M. A. Stoto, and D. Wise. 1986. University-industry research relationships in biotechnology: implications for the university. *Science.* 232(4756):1361-1366.

Bodenheimer, T. 2000. Uneasy alliance: clinical investigators and the pharmaceutical industry. *N. Engl. J. Med.* 342(20):1539-1544.

Bratic, V. W., S. Webster, S. Matthews, and R. S. Harrell. 2005. How patent pools can avoid competition concerns. *Managing Intellectual Property.* 148:44-47.

Bresalier, R. S., R. S. Sandler, H. Quan, J. A. Bolognese, B. Oxenius, K. Horgan, C. Lines, R. Riddell, D. Morton, A. Lanas, M. A. Konstam, J. A. Baron: Adenomatous Polyp Prevention on Vioxx (APPROVe) Trial Investigators. 2005. Cardiovascular events associated with Rofecoxib in a colorectal adenoma chemoprevention trial. *N. Engl. J. Med.* 352(11):1092-1102.

Campbell, E. G., B. R. Clarridge, M. Gokhale, L. Birenbaum, S. Hilgartner, N. A. Holtzman, and D. Blumenthal. 2002. Data withholding in academic genetics: evidence from a national survey. *JAMA*. 287(15):473-480.

Carr, S., R. Aebersold, M. Baldwin, A. Burlingame, K. Clauser, and A. Nesvizhskii. 2004. The need for guidelines in publication of peptide and protein identification data: Working Group on Publication Guidelines for Peptide and Protein Identification Data. *Mol. Cell. Proteomics.* 3: 531-533.

Cho, M. K., S. Illangasekare, M. A. Weaver, D. G. Leonard, and J. F. Merz. 2003. Effects of patents and licenses on the provision of clinical genetic testing services. *J. Mol. Diagn.* 5(1):3-8.

Clark, J., J. Piccolo, B. Stanton, and K. Tyson. 2000. Patent Pools: A Solution to the Problem of Access in Biotechnology Patents. Available at *www.uspto.gov/web/offices/pac/dapp/opla/patentpool.pdf.* Accessed June 21, 2005.

Collins, F. S. 1995. Positional cloning moves from perditional to traditional. *Nat. Genet.* 9(4): 347-350.

Collins, F. 2004. Intellectual property and genomics: a rocky relationship: remarks to Committee on Intellectual Property in Genomic and Protein Research and Innovation at February 11-12 meeting.

Crick, F. 1967. *Of Molecules and Men.* Seattle: University of Washington Press.

Dickson, D. 1993. UK clinical geneticists ask for ban on the patenting of human genes. *Nature.* 366(6454):391.

Dueker, K. S. 1997. Biobusiness on campus: commercialization of university-developed biomedical technologies. *Food Drug Law J.* 52(4):453-509.

Dulbecco, R. 1986. *Mind from Matter? An Essay on Evolutionary Epistemology.* Palo Alto: Blackwell Scientific Publications.

Easton, D. F., D. T. Bishop, D. Ford, and G. P. Crockford. 1993. Genetic linkage analysis in familial breast and ovarian cancer: results from 214 families. The Breast Cancer Linkage Consortium. *Am. J. Hum. Genet.* 52(4):678-701.

Eisenberg, R. S. 1990. Patenting the human genome. *Emory Law J.* 39(3):721-745.

Eisenberg, R. 1996. Intellectual property at the public-private divide: the case of large-scale cDNA sequencing. *Univ. Chi. Law. School Roundtable.* 3:557-573.

Eisenberg, R. S. 1997. Patenting research tools and the law. In: *Intellectual Property Rights and Research Tools in Molecular Biology: Summary of a Workshop Held at the National Academy of Sciences.* Washington, D.C.: National Academy Press.

Eisenberg, R. S. 2001. Bargaining over the transfer of proprietary research tools. In: *Expanding the Boundaries of Intellectual Property*, pp. 223-249, edited by Dreyfuss, R., D. L. Zimmerman, and H. First. Oxford: Oxford University Press.

Federal Trade Commission. 2003. *To promote innovation: The proper balance of competition and patent law and policy.* Washington, D.C.: Federal Trade Commission.

Finnegan, Henderson, Farabow, Garrett & Dunner, LLP. *Biotechnology Innovation Report 2004: Benchmarks.* June 2004.

Firlik, A. D., and D. W. Lowry. 2000. Is academic medicine for sale? *N. Engl. J. Med.* 343(7):509-510.

Gluck, M., D. Blumenthal, and M. Stoto. 1987. University-industry research relationships in the life sciences: Implications for students and post-doctoral fellows. *Research Policy.* 16:327-336.

Graveley, B. R. 2005. Mutually exclusive splicing of the insect Dscam pre-mRNA directed by competing intronic RNA secondary structures. *Cell.* 123(1):65-73.

Gusella, J. F., N. S. Wexler, P. M. Conneally, S. L. Naylor, M. A. Anderson, R. E. Tanzi, P. C. Watkins, K. Ottina, M. R. Wallace, A.Y. Sakaguchi, A. B. Young, I. Shoulson, E. Bonilla, and J. B. Martin. 1983. A polymorphic DNA marker genetically linked to Huntington's disease. *Nature.* 306(5940): 234-238.

Hall J. M., M. K. Lee, B. Newman, J. E. Morrow, L. A. Anderson, B. Huey, and M. C. King. 1990. Linkage of early-onset familial breast cancer to chromosome 17q21. *Science*. 250 (4988):1684-1689.

Heller, M. A., and R S. Eisenberg. 1998. Can patents deter innovation? The anticommons in biomedical research. *Science*. 280(5364):698-701.

Henry, M. R., M. K. Cho, M. A. Weaver, and J. F. Merz. 2002. DNA patenting and licensing. *Science*. 297(5585):1279.

Hollen, T. 2000. NIH researchers receive cut-price BRCA test. *Nat. Med.* 6(6):610.

Howlett, M., and A. F. Christie. 2004. An analysis of the approach of the European, Japanese, and United States patent offices to patenting partial DNA sequences (ESTs). University of Melbourne Faculty of Law, Legal Studies Research Paper No. 82. Available at ssrn.com/abstract=573184.

Hughes, S. S. 2001. Making dollars out of DNA. The first major patent in biotechnology and the commercialization of molecular biology, 1974-1980. *Isis*. 92(3):541-75.

Jensen, K., and F. Murray. 2005. IP Landscape of the Human Genome. *Science*. 310:239-240.

Johnson, R. P., and A. Kaur. 2005. HIV: viral blitzkrieg. *Nature*. 434(7037):1080-1081.

Kastan, M. B., and J. Bartek. 2004. Cell-cycle checkpoints and cancer. *Nature*. 432(7015):316-323.

Kaufman, R. J. 1989. Genetic engineering of factor VIII. *Nature*. 342(6246):207-208.

Kaul R., G. P. Gao, K. Balamurugan, and R. Matalon. 1993. Cloning of the human aspartoacylase cDNA and a common missense mutation in Canavan disease. *Nat. Genet.* 5(2):118-123.

Kendrew, J. C., G. Bodo, H. M. Dintzis, R. G. Parrish, H. Wyckoff, and D. C. Phillips. 1958. A three-dimensional model of the myoglobin molecule obtained by x-ray analysis. *Nature*. 181(4610):662-666.

Kenney, M. 1986. *Biotechnology: The University-Industry Complex*. New Haven: Yale University Press.

Kilpatrick, M. M., D. W. Lowry, A. D. Firlik, H. Yonas, D. W. Marion. 2000. Hyperthermia in the neurosurgical intensive care unit. *Neurosurgery*. 47(4):850-856.

Kodish, E., T. Murray, and P. Whitehouse. 1996. Conflict of interest in university-industry research relationships: realities, politics, and values. *Acad. Med.* 71(12):1287-1290.

Kristiansen, O. P., Z. M. Larsen, and F. Pociot. 2000. CTLA-4 in autoimmune diseases–a general susceptibility gene to autoimmunity? *Genes Immun*. 1(3):170-84.

Lander, E. S., L. M. Linton, B. Birren, C. Nusbaum, M. C. Zody, et al. 2001. Initial sequencing and analysis of the human genome. *Nature*. 409(6822):860-921.

Lawrence, S. 2004. U.S. biopatents decline. *Nat. Biotechnol.* 22(7):797.

MacDonald, M. E., C. M. Ambrose, M. P. Duyao, R. H. Myers, C. Lin, et al. 1993. A novel gene containing a trinucleotide repeat that is expanded and unstable on Huntington's disease chromosomes. The Huntington's Disease Collaborative Research Group. *Cell*. 72(6):971-983.

Merz, J. F. 1999. Disease gene patents: Overcoming unethical constraints on clinical laboratory medicine. *Clin. Chem.* 45(3):324-330.

Merz, J. F., and M. K. Cho. 1998. Disease genes are not patentable: a rebuttal to McGee. *Camb. Q. Healthc. Ethics*. 7:425-428.

Merz, J. F., D. G. Leonard, A. G. Kriss, and M. K. Cho. 2002. Industry opposes genomic legislation. *Nat. Biotechnol.* 20(7):657.

Michelsohn, A. M. 2004. Biotechnology Innovation Report 2004: Benchmarks. Washington, D.C.: Finnegan, Henderson, Farabow, Garrett & Dunner, LLP.

Milstein, C. 1993. Patents on scientific discoveries are unfair and potentially dangerous. *The Scientist*. 7(21):11.

Mowery, David, Frichard Nelson, Bhaven Sampat, and Arvids Ziedonis. 2004. Ivory Tower and Industrial Innovation: University-Industry Technology Transfer Before and After the Bayh-Dole Act. Stanford University Press.

Murray, F., and S. Stern. 2004. *Do Formal Intellectual Property Rights Hinder the Free Flow of Scientific Knowledge? An Empirical Test of the Anti-Commons Hypothesis.* Unpublished manuscript.

National Research Council. Committee on Mapping and Sequencing the Human Genome. 1988. *Mapping and Sequencing the Human Genome.* Washington, D.C.: National Academy Press.

National Research Council. 1997. Intellectual Property Rights and Research Tools in Molecular biology. *Summary of a Workshop Held at the National Academy of Sciences,* February 15-16, 1996. Washington, D.C.: National Academy Press.

National Research Council. Committee on Responsibilities of Authorship in the Biological Sciences. 2003. *Sharing Publication-Related Data and Materials: Responsibilities of Authorship in the Life Sciences.* Washington D.C.: The National Academies Press.

National Research Council. Committee on Intellectual Property Rights in the Knowledge-Based Economy. 2004. *A Patent System for the 21st Century.* Merrill, S. A., R. C. Levin, and M. B. Myers, eds. Washington, D.C.: The National Academies Press.

National Science Foundation. National Science Board. 2004. *Science and Engineering Indicators.* Washington, D.C.: Government Printing Office.

National Institutes of Health. 1998. Report of the National Institutes of Health (NIH) Working Group on Research Tools. Available at *www.nih.gov./news/researchtools/index.htm.* Accessed June 22, 2005.

Oosterwegel, M. A., R. J. Greenwald, D. A. Mandelbrot, R. B. Lorsbach, and A. H. Sharpe. 1999. CTLA-4 and T cell activation. *Curr. Opin. Immunol.* 11(3):294-300.

Organisation for Economic Co-operation and Development. 2002. *Genetic Inventions, Intellectual Property Rights and Licensing Practices,* p. 48. Available at *www.oecd.org/dataoecd/42/21/2491084.pdf.* Accessed June 22, 2005.

Organisation for Economic Co-operation and Development. 2005. *Intellectual Property and Competition Policy in the Biotechnology Industry.* Available at *www.oecd.org/dataoecd/36/4/35040373.pdf.* Accessed June 22, 2005.

Patent and Trademark Act Amendments of 1980 (Bayh-Dole Act), Pub. L. No. 96-517, 94 Stat. 3015 (codified in scattered sections of 35 U.S.C.).

Petricoin, E. F., and L. A. Liotaa. 2002. Proteomic analysis at the bedside: Early detection of cancer. *Trends Biotechnol.* 20(12 Suppl):S30-4.

Pressman, L., R. Burgess, R. M. Cook-Deegan, S. J. McCormack, I. Nami-Wolk, M. Soucy, and L. Walters. 2005. Patenting and licensing practices for DNA-based patents at U.S. academic institutions. *Nat. Biotechnol.* In press.

Principles and Guidelines for Recipients of NIH Research Grants and Contracts on Obtaining and Disseminating Biomedical Research Resources: Final Notice, 64 *Fed. Reg.* 72,090 (proposed Dec. 23, 1999).

Rosenshine, I., T. Umanski, O. Ilan, A. Peleg-Lavi, Y. Fang, I. Nisan, and D. Friedberg 1999. *Regulation of Virulence Genes and Host Specificity by EPEC.* Presented at UNESCO Workshop on Science for Peace and Regional Scientific Cooperation in Molecular Biology, Microbiology, and Biotechnology. Jerusalem, Israel.

Sampat, B. *Patents on Academic Genomic Discoveries: Effects on Biomedical Research.* Working Paper, Department of Health Management and Policy, University of Michigan. Forthcoming.

Sanger, F., S. Nicklen, and A. R. Coulson. 1977. DNA sequencing with chain-terminating inhibitors. *Proc. Natl. Acad. Sci.* 74(12):5463-5647.

Service, R. F. 2001. Proteomics: Gene and protein patents get ready to go head to head. *Science.* 294(5549):2082-2083.

Sheridan, C. 2004. Curie's victory over BRCA1. *Nat. Biotechnol.* 22(7):797.

Shimbo, I., R. Nakajima, S. Yokohama, and K. Sumikura. 2004. Patent protection for protein structure analysis. *Nat. Biotechnol.* 22(1):109-112.

Staudt, L. M., and S. Dave. 2005. The biology of human lymphoid malignancies revealed by gene expression profiling. *Adv. Immunol.* 87:163-208.

Thackray, A., ed. 1998. *Private Science: Biotechnology and the Rise of the Molecular Sciences.* Philadelphia: University of Pennsylvania Press.

Thursby, M. C., J. Thursby, and E. Dechenaux. 2005. *Shirking Sharing Risk, and Shelving: The role of University License Contracts.* (Working Paper No. 11128). National Bureau of Economic Research, Cambridge, MA.

U.S. Congress, Office of Technology Assessment. 1988. *Mapping Our Genes. Genome Projects: How Big? How Fast? OTA-BA-273.* Washington, D.C.: U.S. Government Printing Office.

Venter, J. C., M. D. Adams, E. W. Myers, P. W. Li, R. J. Mural, et al. 2001. The sequence of the human genome. *Science.* 291(5507):1304-1351.

Vermij, P. 2005. BRAC1 patent revisited. *Nat. Biotechnol.* 23(3):277.

Vinarov, S. D. 2003. Patent protection for structural genomics-related inventions. *J. Struct. Funct. Genomics.* 4(2-3):191-209.

Walsh, J. P., A. Arora, and W. M. Cohen. 2003. Research tool patent and licensing and biomedical innovation. In: *Patents in the Knowledge-Based Economy*, pp. 285-340. W. Cohen and S. Merrill, eds. Washington, D.C.: The National Academies Press.

Walsh, J. P., C. Cho, and W. M. Cohen. 2005. Patents, Material Transfers, and Access to Research Inputs in Biomedical Research at *http://www.uic.edu/~jwalsh/NASreport. html.*

Watson, J. D., and F. H. Crick. 1953. Molecular structure of nucleic acids; a structure for deoxyribose nucleic acid. *Nature.* 171(4356):737-738.

Watson F. L., R. Puttmann-Holgado, F. Thomas, D. L. Lamar, M. Hughes, M. Kondo, V. I. Rebel, and D. Schmucker. 2005. Extensive diversity of Ig-superfamily proteins in the immune system of insects. *Science.* 309(5742):1874-1878.

Wright, B., and P. G. Pardey. 2005. Intellectual property protection: A challenge to agricultural biotechnology innovation developing countries. Unpublished paper.

Yamamoto, Y., and R. B. Gaynor. 2001. Role of the NF-kappaB pathway in the pathogenesis of human disease states. *Curr. Mol. Med.* 1(3):287-296.

Appendixes

Appendix A

Biographical Information of Committee and Staff

Shirley M. Tilghman (*Co-Chair*) has served as president of Princeton University since June 2001. A world-renowned scholar in the field of molecular biology, she served on the Princeton faculty for 15 years before being named president. A native of Canada, Dr. Tilghman received her Honors B.Sc. in chemistry from Queen's University in Kingston, Ontario, in 1968 and her Ph.D. in biochemistry from Temple University in Philadelphia. She came to Princeton in 1986 as the Howard A. Prior Professor of the Life Sciences. Two years later, she joined the Howard Hughes Medical Institute as an investigator. In 1998, she took on additional responsibilities as the founding director of Princeton's multidisciplinary Lewis-Sigler Institute for Integrative Genomics. A member of the National Research Council's committee that set the blueprint for the U.S. effort in the Human Genome Project, Dr. Tilghman also was one of the founding members of the National Advisory Council of the Human Genome Project Initiative for the National Institutes of Health. She is a member of the National Academy of Sciences, the Institute of Medicine, the American Philosophical Society, and the Royal Society of London.

Hon. Roderick R. McKelvie (*Co-Chair*) is a partner in the law firm of Covington & Burling. From March of 1992 to June of 2002, he was a United States District Court Judge for the District of Delaware. During his 10 years on the bench, Judge McKelvie handled a number of patent infringement cases and has written and spoken extensively on issues relating to intellectual property. He is a professorial lecturer in law at George Washington University School of Law and teaches a course in patent enforcement. He is currently president of the Giles Sutherland Rich American Inn of Court. He participated in the inaugural conference of the

STEP Board project, Intellectual Property in the Knowledge-Based Economy. He has a degree in economics from Harvard University and a J.D. from the University of Pennsylvania.

Ashish Arora is professor of Economics and Public Policy at Carnegie Mellon University, with a courtesy appointment in the School of Computer Science. Professor Arora's research centers on the economics of technological change, including topics such as intellectual property rights and technology licensing. He also has published extensively on the software, biotechnology, and chemical industries. An enduring research interest is understanding the rise and functioning of markets for technology and their consequences for strategy, industry structure, and economic development. Professor Arora co-directs the Software Industry Center at Carnegie Mellon University. He earned his doctorate in economics from Stanford University in 1992 with a dissertation titled *Technology Licensing, Tacit Knowledge, and the Acquisition of Technological Capability.*

Helen M. Berman is Board of Governors Professor of Chemistry and Chemical Biology at Rutgers University. Her research area is structural biology and bioinformatics with a special focus on protein nucleic acid interactions. She is the founder of the Nucleic Acid Database, a repository of information about the structures of nucleic acid containing molecules, and is the co-founder and director of the Protein Data Bank, which is the international repository of the structures of biological macromolecules. She is a fellow of the American Association for the Advancement of Science and the Biophysical Society, from which she received the Distinguished Service Award in 2000. She is also the past president of the American Crystallographic Association. Dr. Berman received her A.B. in 1964 from Barnard College and a Ph.D., 1967, from the University of Pittsburgh.

Joyce Brinton recently retired as the director of Harvard University's Office for Technology and Trademark Licensing, a position she had held since 1984. Before assuming this position, she spent six years in the Dean's Office at Harvard Medical School, where she first became involved in technology transfer. Before that she was the asssistant director for Administration of a laboratory at the Medical School. She has been with Harvard University for 37 years. Her prior experience included work with a research foundation and with a pharmaceutical company. Ms. Brinton received her degree in Biology from Washington University in St. Louis. She is actively involved at the national level in developing university technology transfer policy. She is a past president of AUTM (the technology transfer professional organization) and is a recipient of its Bayh-Dole Award in recognition of her contributions to the field.

Stephen Burley is the chief scientific officer of Structural GenomiX, Inc. (SGX), located in San Diego, California. SGX is an oncology-focused drug discovery and development company, which currently has Troxatyl® in clinical trials and multiple protein kinase inhibitors in preclinical development, including imatinib-resistant BCR-ABL. Prior to joining SGX, Burley was the Richard M. and Isabel P. Furlaud Professor and Chief Academic Officer at the Rockefeller University and a full investigator of the Howard Hughes Medical Institute. He is a fellow of the Royal Society of Canada and of the New York Academy of Sciences. Burley received an M.D. degree from Harvard Medical School in the joint Harvard-MIT Health Sciences and Technology program and, as a Rhodes Scholar, received a D.Phil. in Molecular Biophysics from Oxford University. He trained in internal medicine at the Brigham and Women's Hospital and did postdoctoral work with William N. Lipscomb at Harvard University and Gregory A. Petsko at the Massachusetts Institute of Technology. With William J. Rutter and others at the University of California, San Francisco, and the Rockefeller University, Burley co-founded Prospect Genomics, Inc., which was subsequently acquired by SGX.

Q. Todd Dickinson is Vice President and Chief Intellectual Property Counsel, General Electric Co., where he has corporate-wide responsibility for all intellectual property and technology licensing matters. Mr. Dickinson previously served under President Clinton as Under Secretary of Commerce for Intellectual Property and Director of the U.S. Patent and Trademark Office. Prior to assuming his present position, he was a partner in the law firm of Howrey Simon Arnold & White, where he was a leader of its intellectual property practice. Mr. Dickinson has experience in all aspects of intellectual property law and public policy, including patents, trademarks, copyrights, and trade secrets. He has written and spoken extensively on intellectual property issues, and has testified before Congress, the Federal Trade Commission and the National Academy of Sciences on the impact of intellectual property law and policy. Mr. Dickinson is a member of the Board of Directors of the Intellectual Property Owners Association and the Council of the American Bar Association Intellectual Property Law Section. He is also the ABA delegate to the World Intellectual Property Organization (WIPO). In 2004 and 2005, he was named one of "The 50 Most Influential People in Intellectual Property" by *Managing Intellectual Property* magazine. He has also taught or lectured at various universities, including Stanford, Yale, University of California (Berkeley), MIT, Georgetown, George Washington and Tokyo University, and is a member of the Editorial Advisory Board of the BNA *Patent, Trademark & Copyright Journal*. Mr. Dickinson earned his J.D. in 1977 from the University of Pittsburgh and his B.S. from Allegheny College in 1974. He is admitted to the bars of the District of Columbia, Pennsylvania, Illinois and California, the United States Patent and Trademark Office, and the Court of Appeals for the Federal Circuit.

Rochelle Dreyfuss is the Pauline Newman Professor of Law at New York University School of Law. After spending several years as a research chemist at

Vanderbilt University Medical School, the Albert Einstein Medical School, and the Ciba Geigy Corporation, Dreyfuss entered Columbia University Law School, where she was Articles and Book Review Editor of the *Columbia Law Review*. Following her graduation in 1981, she became law clerk first to Chief Judge Wilfred Feinberg of the United States Court of Appeals for the Second Circuit and later to Chief Justice Warren E. Burger of the Supreme Court. In 1983, Dreyfuss began teaching at NYU. Her research and teaching interests include intellectual property, privacy, the relationship between science and law, and civil procedure. She has authored many articles on these subjects and has co-authored casebooks on civil procedure and intellectual property law. Previously a consultant to the Federal Trade Commission, the Federal Courts Study Committee, and the Presidential Commission on Catastrophic Nuclear Accidents, she also has served on the National Academies' Committee on Intellectual Property Rights in the Knowledge-Based Economy (chaired by Richard Levin, President, Yale University), as well as on the Science and Law and Patent Law Committees of the Association of the Bar of the City of New York. Professor Dreyfuss is currently a co-reporter of the American Law Institute's Project on Intellectual Property: Principles Governing Jurisdiction, Choice of Law, and Judgments in Transnational Disputes. Her undergraduate degree is from Wellesley College. She also holds an M.S. in Chemistry from the University of California, Berkeley, and is admitted to practice in New York. She has visited at the law schools of the University of Chicago, the University of Washington, and Santa Clara and has lectured all over the world.

Rebecca S. Eisenberg, J.D., is Robert and Barbara Luciano Professor of Law at the University of Michigan in Ann Arbor, where she teaches courses in patent law, trademark law, and drug regulation and runs an intellectual property workshop. She has previously taught courses in tort law, legal regulation of science, and legal issues associated with the Human Genome Project. She has written and lectured extensively about patent law as applied to biotechnology and the role of intellectual property at the public-private divide in research science, publishing in scientific and medical journals as well as in law reviews. She spent the 1999-2000 academic year as a visiting professor of law, science, and technology at Stanford Law School. She has received grants from the program on Ethical, Legal, and Social Implications of the Human Genome Project of the U.S. Department of Energy Office of Biological and Environmental Research for her work on private appropriation and public dissemination of DNA sequence information. Professor Eisenberg has played an active role in public policy debates concerning the role of intellectual property in biomedical research. She received a Distinguished Service Award from the Berkeley Center for Law & Technology in 2002. She serves as a member of the Science, Technology, and Law Panel.

Charles M. Hartman (deceased) was a general partner with CW Group, a leading manager of medical venture capital funds focused on seed and early stage health care investing located in New York. Mr. Hartman is a chemistry graduate of the University of Notre Dame with an M.B.A. degree from the University of Chicago. His experience, prior to joining the CW Group, included 17 years at Johnson & Johnson. At the CW Group, his areas of specialization included genomics, pharmaceuticals, and information systems. Examples of Mr. Hartman's earlier start-ups include Athena Neurosciences (initially went public, then was acquired by Elan Corporation) and Sugen, Inc. (initially went public then was acquired by Pharmacia), while more recent examples include Surface Logix and Plexxikon.

Daniel J. Kevles, B.A. (Physics); Princeton University; Oxford University (European History); Ph.D. (History), Princeton University, is Stanley Woodward Professor of History, Yale University, and the J.O. and Juliette Koepfli Professor of the Humanities Emeritus at the California Institute of Technology. His research interests and extensive writing include the interplay of science and society past and present; the history of science in America; the history of modern physics; the history of modern biology; the Human Genome Project; and scientific fraud and misconduct. He is currently working on a book on the history of intellectual property in living organisms, a subject that he teaches in the Yale Law School. He is author of *The Baltimore Case: A Trial of Politics, Science, and Character*; *In the Name of Eugenics: Genetics and the Uses of Human Heredity*; and *The Physicists: The History of a Scientific Community in Modern America.* He is co-editor of *The Code of Codes: Scientific and Social Issues in the Human Genome Project,* and a co-author of *Inventing America: A History of the United States*. His past service on National Academies' committees includes (1) the Committee for "Innovations in Computing and Communications: Lessons from History"; (2) the U.S. National Committee for the International Union of the History and Philosophy of Science (chair); (3) the Delegation to the General Assembly and 18th International Congress of History of Science, Hamburg/Munich, Germany (chair); (4) the U.S. National Committee for the International Union of the History and Philosophy of Science (member and vice chair); (5) the National Cancer Policy Board; and (6) the Committee on Large-Scale Science and Cancer Research.

David Korn is senior vice president for Biomedical and Health Sciences Research at the Association of American Medical Colleges (AAMC) in Washington, D.C., a position he assumed in 1997. Prior to that, Dr. Korn had spent 29 years as Professor of Pathology at Stanford University, where he was chairman of the Department of Pathology from 1967 to 1984, and then University Vice President and Dean of the Medical School from 1984 to 1995. Dr. Korn is a Member of the Institute of Medicine, and a Fellow of the AAAS, where he formerly served on the Council. From 1998 to 2004 he was a member of the University Grants Committee of Hong Kong, where he served as chairman of the Medical Subcommittee. Dr. Korn has been a member of the editorial boards of the *American Journal of Pathology*, *The Journal of Biological Chemistry*, and *Human Pathology*. He has been a member of many Societies, Councils, and Boards, and from 1984 to 1991, he was chairman of the National Cancer Advisory Board. He has written numerous scientific articles, earlier in his career in bacteriophage genetics and the biochemistry and molecular biology of DNA replication in human cells, and more recently about issues of health and science policy, topics in which he has been heavily engaged on the national scene. Dr. Korn has previously served on the National Academies' (1) Committee on EPA Assessment Factors for Data Quality; (2) Committee on Interactions of Drugs, Biologics, and Chemicals in U.S. Military Forces; and (3) Board of Life Sciences: Clinical Research Roundtable (of which he was a founder). He currently serves on the Committee on Patenting in Human Genomics and Proteomics. Dr. Korn is a member of the Science, Technology, and Law Panel.

George M. Milne, Jr., is a venture partner at Radius Ventures, where he takes an active role in selecting and guiding new investments in the biomedical arena. He is Chairman of Phylogix, Inc., and serves on the corporate boards of Mettler-Toledo, Inc., Charles River Laboratories, Inc., MedImmune, Inc., Conor Medsystems, Inc., Aspreva Pharmaceuticals, Rib-X Pharmaceuticals and Athersys, Inc. Dr. Milne is a member of the Scientific Advisory Board of MedImmune, Inc. In 2002, he retired from Pfizer Inc., after 32 years, where he served as a corporate senior vice-president and as president of Central Research, with global responsibility for Human and Veterinary Medicine Research and Development from 1993 to 2000. During his tenure as president, Dr. Milne led the unprecedented growth of the Research Division as evidenced by a five-fold increase in the annual investment, a more than doubling of the scientific staff, and the creation of an external alliance portfolio with an investment of more than $228 million. Following the acquisition of Warner-Lambert, Dr. Milne assumed the role of executive vice president of Pfizer global research and development and president of strategic and operations management. Dr. Milne's interests include the community, a diversity of intellectual pursuits and the public policy arena. He is an adjunct senior lecturer in the Harvard-MIT Division of Health Sciences Technology. He chairs the Board of the Mystic Aquarium and the Institute for Exploration and is active on a number of other boards, including the New York Botanical Garden and the National Foundation for Infectious Diseases. Dr Milne received a B.Sc. in chemistry from Yale University in 1965. He earned his Ph.D. from the Massachusetts Institute of Technology in 1969 and completed postdoctoral work at Stanford University. In 1975 he attended the Medical College of Virginia for postgraduate training in pharmacology.

Richard Scheller is the executive vice president, Research, at Genentech and an Adjunct Professor, Department of Biochemistry & Biophysics, School of Medicine, University of California, San Francisco. He is responsible for setting the strategy for Genentech's research and drug discovery activities and is a member of Genentech's Executive Committee. He is a world-renowned cell biologist and has defined the molecular mechanisms of exocytosis, particularly as it pertains to neurotransmitter release. He has also contributed to the general understanding of the molecular mechanisms that regulate membrane organization and transport in eukaryotic cells. He received the Alan T. Waterman Award from the National Science Foundation and the National Academy of Sciences Award in Molecular Biology. Dr. Scheller is a member of the National Academy of Sciences and a Fellow of the American Academy of Arts and Sciences. He received his B.S. degree from the University of Wisconsin, Madison, Department of Biochemistry and his Ph.D. from the California Institute of Technology, Division of Chemistry.

Rochelle Seide is a partner in the law firm of Arent Fox PLLC. Dr. Seide has been practicing intellectual property law for nearly 20 years, primarily in the life sciences. Dr. Seide received a B.S. in bacteriology and botany from Syracuse University, an M.S. in biology (immunology) from Long Island University, a Ph.D. in Human Genetics from the City University of New York, Mt. Sinai School of Medicine, and a J.D. from the University of Akron School of Law. Prior to entering the legal profession, she served on the faculty of a medical school, where she taught and carried out research in medical genetics, microbiology, and immunology. She has obtained patents in the areas of biotechnology, chemistry, and pharmaceuticals for a variety of clients. Dr. Seide also counsels clients on legal issues relating to biotechnology and pharmaceutical patents, including patent enforcement, validity, and infringement, licensing, due diligence, and business development. She has experience in transactional matters for biotechnology and pharmaceutical clients. She has litigated patent matters before the federal courts, as well as represented clients in *inter parties* patent interferences in the U.S. Patent and Trademark Office (USPTO) in biotechnology, pharmaceutical, and medical device technologies. She also has served as a patent law expert in patent litigation relating to recombinant hormones, starch-based biodegradable plastics, assays for genetic polymorphisms, and treatment for bacterial sepsis. Dr. Seide is a frequent speaker and author on a variety of life sciences intellectual property issues and has also taught a seminar course on biotechnology patent law at the University of Akron School of Law. She is a member of a number of intellectual property law associations, including the American Intellectual Property Law Association, where she served as chair of the Biotechnology Committee, and the New York Intellectual Property Law Association. Recognized as an expert in her field, she is listed in the Chambers USA Guide *America's Leading Business Lawyers (2004-2005)*, Chambers Global *The World's Leading Lawyers (2004-2005)*, and the *Best Lawyers in America 2005*. She is admitted to the bars of Ohio, New

York, several U.S. District Courts, and the Court of Appeals for the Federal Circuit, and she is registered to practice before USPTO. Dr. Seide was previously a partner at Baker Botts LLP and at Brumbaugh, Graves, Donohue & Raymond.

Robert H. Waterston, M.D., Ph.D., is the William H. Gates III Endowed Chair in Biomedical Sciences and chair of the Department of Genome Sciences at the University of Washington, Seattle. In his role as director of the Genome Sequencing Center at Washington University School of Medicine in St. Louis, Waterston brought whole-genome sequencing of metazoan organisms to reality. In collaboration with the Sanger Centre, he constructed a physical map and obtained the complete sequence of the first animal genome, the nematode *Caenorhabditis elegans*. His laboratory contributed to the completion of the *S. cerevesiae*, *A. thaliana*, and other genomes and was a principal partner in the Human Genome Project, constructing the clone map and contributing 20 percent of the sequence of the human genome to the effort. He also pioneered the use of the Internet for the rapid release of sequence and map information. His contributions to large-scale DNA sequencing remain central to the ongoing analysis of the human genome. He received his B.S.E. from Princeton and his M.D. and Ph.D. from the University of Chicago. He is a member of the National Academy of Sciences, the Institute of Medicine, and the American Academy of Arts and Sciences. He was awarded the Dan David Prize, the Gairdner Award, and the General Motors Research Foundation Sloan Award, among others.

Nancy Wexler, Ph.D., is Higgins Professor of Neuropsychology in the Departments of Neurology and Psychiatry of the College of Physicians and Surgeons of Columbia University. She also serves as president of the Hereditary Disease Foundation, founded in 1968 by her father and dedicated to curing Huntington's disease (HD). She received her B.A. cum laude from Radcliffe College and her Ph.D. in psychology from the University of Michigan. She served as executive director of the Congressional Commission for the Control of Huntington's Disease and its Consequences and then worked for the National Institute of Neurological Disorders and Stroke, National Institutes of Health (NIH). Dr. Wexler has led a continuing 25-year study in Venezuela focusing on the largest family worldwide with HD. Her work was critical to the discovery of the gene for HD, localized in 1983 and isolated in 1993. She was involved in developing guidelines for delivering HD presymptomatic genetic testing. Interest in issues raised by genetic testing led her to be selected as chair of the Joint NIH/DOE Ethical, Legal, and Social Issues Working Group of the National Center for Human Genome Research and co-chair of the Ethical, Legal, and Social Issues Committee of the Human Genome Organization. She was a member of the first Program Advisory Committee of the National Human Genome Research Institute, NIH. Dr. Wexler has served as a member of the board of directors of the American Association for the Advancement of Science and of the Advisory Committee on Research on

Women's Health, NIH. Dr. Wexler was elected a fellow of the Royal College of Physicians; a fellow of the American Association for the Advancement of Science, Section on Neuroscience; a Member of the European Academy of Sciences and Arts; and Councilor, Society for Neuroscience. She is an honorary Fellow of the New York Academy of Sciences and a Member of the Institute of Medicine. Notable awards include a Fulbright Scholarship, three honorary doctorates, the first Robert J. and Claire Pasarow Foundation Award, the Foster Elting Bennett Memorial Lecture, the J. Allyn Taylor International Prize in Medicine, and the Albert Lasker Public Service Award.

Brian Wright is professor of Agricultural and Resource Economics at the University of California, Berkeley. He received a Bachelor of Agricultural Economics (First Class Honors and University Medal) from the University of New England, Armidale, Australia, and was awarded one of the two Frank Knox Fellowships given annually to Australian students by Harvard University, where he received an A.M. and a Ph.D. in economics. He then joined the Economics Department at Yale University in 1975, and moved to Berkeley in 1985. He is a fellow of the American Agricultural Economics Association. His research interests include the following: the economics of research and development, project evaluation, research incentives including patents, prizes and contractual arrangements, intellectual property rights and their licensing, the economics of conservation of genetic resources, agricultural policy, and the economics of markets for storable commodities, market stabilization, energy markets, and insurance. He publishes in economics, agricultural economics, and in scientific publications. His books include *Storage and Commodity Markets* (with Jeffrey Williams), *Reforming Agricultural Commodity Policy* (with Bruce Gardner), and *Saving Seeds*, with Bonwoo Koo and Philip Pardey and others. He served as the economist member of the Subcommittee on Proprietary Science and Technology of the Consultative Group on International Agricultural Research (CGIAR) and has advised the International Plant Genetic Resources Institute on the economics of conservation of genetic resources for the CGIAR. He has consulted for the United States Department of Justice, the Food and Agriculture Organization of the United Nations, the World Bank, and the International Food Policy Research Institute, and has served as an expert witness in litigation regarding patent licensing and agricultural biotechnology. His work evaluating the cost of international *ex situ* conservation of agricultural germplasm helped establish the basis for the Global Crop Diversity Trust, which is raising $260 million for the secure and sustainable funding of global crop genetic resources for the long term. He is an advisor for an initiative on prizes for innovation in African agriculture, is involved in the development of the Biological Innovation for Open Society initiative, and has been involved directly, and through his students, in the development of the Public Intellectual Property for Agriculture initiative.

STAFF

Stephen A. Merrill has been executive director of the National Academies' Board on Science, Technology, and Economic Policy (STEP) since its formation in 1991. The STEP program addresses macroeconomic, intellectual property, technical standards, trade, taxation, human resources, and statistical as well as research and development policies affecting technology development and economic performance. Dr. Merrill has directed several STEP projects and publications, including *Investing for Productivity and Prosperity* (1994); *Improving America's Schools* (1995); *Industrial Research and Innovation Indicators* (1997); *U.S. Industry in 2000: Studies in Competitive Performance* and *Securing America's Industrial Strength* (1999); *Trends in Federal Support of Research and Graduate Education* (2001); and *A Patent System for the 21st Century* (2004). For his work on *A Patent System for the 21st Century*, he was named one of the 50 most influential people worldwide in the intellectual property field by *Managing Intellectual Property* magazine. Dr. Merrill's association with the National Academies began in 1985, when he was principal consultant on the Academy report, *Balancing the National Interest: National Security Export Controls and Global Economic Competition.* In 1987 he was appointed to direct the Academies' first government and congressional liaison office. During his tenure as executive director of Government and External Affairs, the Academies received a steadily increasing number of congressional requests for policy advice. Previously, Dr. Merrill was a fellow in International Business at the Center for Strategic and International Studies, where he specialized in technology trade issues. For seven years until 1981, he served on various congressional staffs, most recently that of the Senate Commerce, Science, and Transportation Committee, where he organized the first congressional hearings on international competition in biotechnology and microelectronics and was responsible for legislation on technological innovation and the allocation of intellectual property rights arising from government-sponsored research. Dr. Merrill holds degrees in political science from Columbia (B.A., summa cum laude), Oxford (M. Phil.), and Yale (M.A. and Ph.D.) From 1989 to 1996 he was an adjunct professor of international affairs at Georgetown University.

Anne-Marie Mazza, B.A., Economics; M.A., History and Public Policy; Ph.D., Public Policy, The George Washington University, joined the National Academies in 1995. She has served as senior program officer with both the Committee on Science, Engineering, and Public Policy and the Government-University-Industry Research Roundtable. In 1999, she was named the first director of the Committee on Science, Technology, and Law, an activity designed to foster communication and analysis among scientists, engineers, and members of the legal community. Dr. Mazza has been the study director on numerous Academies' reports, including *Intentional Human Dosing Studies for EPA Regulatory Purposes:*

Scientific and Ethical Issues (2004); *Ensuring the Quality of Data Disseminated by the Federal Government* (2003); *The Age of Expert Testimony: Science in the Courtroom* (2002); *Issues for Science and Engineering Researchers in the Digital Age* (2001); and *Observations on the President's Fiscal Year 2000 Federal Science and Technology Budget* (1999). Between October 1999 and October 2000, she divided her time between the Academies and the White House Office of Science and Technology Policy, where she served as a senior policy analyst responsible for issues associated with a Presidential Review Directive on the government-university research partnership. Before joining the Academies, Dr. Mazza was a senior consultant with Resource Planning Corporation.

Craig Schultz has been with the National Academies' Board on Science, Technology, and Economic Policy (STEP) since 1998. He has worked on several STEP projects on human resources, government-industry partnerships, research and development, and intellectual property rights. Prior to joining STEP, Mr. Schultz worked in the Office of the Vice President for Development at the University of Virginia. He holds a B.A., High Honors, from the University of Michigan and an M.A. from the University of Virginia.

Stacey Speer, B.S., Biomedical Engineering, University of Tennessee, joined the National Academies' Science, Technology, and Law Program in September 2002 as the Christine Mirzayan Intern. Ms. Speer is now the senior program assistant of the Science, Technology, and Law Program. She is attending the George Washington University, pursuing a master's of Forensic Science.

Patricia E. Santos, M.Ed., is the program associate with the Science, Technology, and Law (STL) Panel. Before joining STL in April 2005, she worked in the Board of Higher Education and Workforce Unit at the National Academies on National Institutes of Health training assessment studies. Prior to coming to the Academies, she taught middle school math and received the Maryland State Governor's Award for Excellence in Math, Science, and Technology instruction.

Kathi E. Hanna, M.S., Ph.D., is a science and health policy consultant, writer, and editor specializing in biomedical research policy and bioethics. She served as Research Director and Senior Consultant to President Clinton's National Bioethics Advisory Commission and as Senior Advisor to President Clinton's Advisory Committee on Gulf War Veterans' Illnesses. More recently, she served as the lead author and editor of President Bush's Task Force to Improve Health Care Delivery for Our Nation's Veterans. In the 1980s and 1990s, Hanna was a senior analyst at the congressional Office of Technology Assessment, contributing to numerous science policy studies requested by congressional committees on science education, research funding, biotechnology, women's health, human genetics, bioethics, and reproductive technologies. In the past decade, she has served as an analyst and editorial consultant to the Howard Hughes Medical Institute, the National Institutes of Health, the Institute of Medicine, the National Academies, and to several charitable foundations, voluntary health organizations, and biotechnology companies. Before coming to Washington, D.C., she was the genetics coordinator at Children's Memorial Hospital in Chicago, where she directed clinical counseling and coordinated an international research program in prenatal diagnosis. Hanna received an A.B. in biology from Lafayette College, an M.S. in human genetics from Sarah Lawrence College, and a Ph.D. from the School of Business and Public Management, George Washington University.

Sara Davidson Maddox, M.A., is a science and health policy writer and editor, with extensive experience in the areas of bioethics, biomedical research, and health services and quality. She was editor for the National Bioethics Advisory Commission and has participated in projects for the National Institutes of Health and the Institute of Medicine.

Appendix B

Search Algorithms Used to Identify Patents of Interest

The following search algorithm is used by the Georgetown University research team to identify DNA-based U.S. patents. The search is in the patent "claims" field.

((047???* OR 119* OR 260???* OR 426* OR 435* OR 514* OR 536022* OR 5360231 OR 536024* OR 536025* OR 800*) <in> NC) AND (("antisense" OR <case><wildcard>cDNA* OR centromere OR deoxyoligonucleotide OR deoxyribonucleic OR deoxyribonucleotide OR <case><wildcard>DNA* OR exon OR "gene" OR "genes" OR genetic OR genome OR genomic OR genotype OR haplotype OR intron OR <case><wildcard>mtDNA* OR nucleic OR nucleotide OR oligonucleotide OR oligodeoxynucleotide OR oligoribonucleotide OR plasmid OR polymorphism OR polynucleotide OR polyribonucleotide OR ribonucleotide OR ribonucleic OR "recombinant DNA" OR <case><wildcard>RNA* OR <case><wildcard>mRNA* OR <case><wildcard>rRNA* OR <case><wildcard>siRNA* OR <case><wildcard>snRNA* OR <case><wildcard>tRNA* OR ribonucleoprotein OR <case><wildcard>hnRNP* OR <case><wildcard>snRNP* OR <case><wildcard>SNP*) <in> CLAIMS))

The following search algorithms were used by Academies staff to identify U.S. patents in several genomic and proteomic categories, molecular pathways, and research tools. In all but one case the patent "claims" field was searched. The NF-kB pathway was searched in the "keyword" field because it includes an assignee restriction in the Boolean string:

1. **Genes and gene regulatory sequences:** ("nucleic acid" OR nucleotide OR "nucleotide sequence" OR oligonucleotide OR deoxyribonucleic OR deoxyribonucleotide OR oligoribonucleotide OR ribonucleotide OR "recombinant DNA" OR cDNA OR plasmid OR gene OR genomic) AND ("promoter" OR "enhancer" OR "response element" OR "DNA motif" OR "DNA binding" OR "upstream region")

2. **SNPs and/or haplotypes:** (Haplotype OR Polymorphism OR "single nucleotide polymorphism" OR "variable number of tandem repeat polymorphisms" OR "tandem repeats" OR "microsatellite polymorphisms" OR allele OR "genotypic variation" OR "genetic locus" OR "DNA polymorphism" OR "restriction fragment length polymorphism")

3. **Gene expression profiles/profiling:** ("nucleic acid" OR nucleotide OR "nucleotide sequence" OR oligonucleotide OR deoxyribonucleotide OR "recombinant DNA" OR cDNA OR plasmid) AND ("gene expression profile" OR detection OR array OR screen OR "microarray" OR diagnostic OR treatment)

With U.S. class restriction: 800* OR 435* OR 424* OR 535* OR 935* OR 530* OR 514* OR 436*

4. **Protein structure:** (protein OR polypeptide OR oligopeptide or proteome OR protease OR enzymatic OR "enzymatic polypeptide" OR peptide OR "protein complex" OR "protein domain" OR PDB OR "protein data bank" OR motif OR antibody OR antibodies or enzyme) AND ("three-dimensional structure" OR angstrom OR "atomic coordinate" OR coordinate OR "space group" OR "binding pocket" OR "binding domain" OR "fold space" OR "modeling test compounds") AND ("mass spectroscopy" OR MS OR "mass spectrometry" OR crystallography OR crystallographic OR NMR OR "nuclear magnetic resonance" OR "x-ray crystallography" OR "crystal structure" OR "computational modeling" OR "computer readable storage medium" OR algorithm OR "crystalline form" OR "in silico screening")

5. **Protein-protein interactions:** (protein OR polypeptide OR oligopeptide OR peptide OR proteome OR protease OR enzymatic OR "enzymatic polypeptide" OR peptide OR "protein complex" OR "protein domain" OR PDB OR "protein data bank" OR motif OR antibody OR antibodies OR enzyme OR factor OR homolog OR homologue OR analog OR analogue OR ortholog OR orthologue) AND ("interaction partner" OR (("protein-protein" OR "protein-DNA" OR "DNA-protein") AND (binding or interaction or assembly)) OR "receptor-ligand" OR ((binding OR interaction OR interacting OR active) AND (domain OR site OR region OR pocket)) OR "receptor-agonist" OR "receptor/agonist" OR "re-

ceptor-antagonist" OR "receptor/antagonist" OR "receptor-target" OR "receptor/target" OR bivalent OR "agonist-antagonist" OR "agonist/antagonist")

With U.S. class restriction: 800* OR 435* OR 424* OR 536* OR 935* OR 530* OR 514* OR 436*

6. **Modified animals:** ((Transgenic or "targeted deletion" or "targeted ablation" or knockout) NOT plant)

With U.S. class restriction: 800* OR 435* OR 424* OR 536* OR 935* OR 530* OR 514* OR 436*

7. **Software:** Software and protein OR software and genetics OR software and "nucleic acid" OR software and "systems biology" OR software and "protein regulation pathways" OR software and "protein regulation pathways" OR "evolutionary computation" and software OR "genetic programming" and software

With U.S. class restriction: 435

8. **Algorithms:** Algorithms and genetic OR algorithms and protein OR algorithms and haplotype OR algorithms and biological evolution OR "evolutionary computation" OR "genetic algorithms" OR "genetic programming" OR "modeling genetic inheritance" OR "biological evolution" and modeling OR "medical informatics" OR "sequencing algorithms" OR "informatics" and protein

With U.S. class restriction: 435